Young Einstein

Other Books by Barry Parker

Einstein's Dream: The Search for Unified Theory
Albert Einstein's Vision
Einstein's Brainchild: Relativity Made Rel. Easy
Einstein: The Passions of a Scientist
Good Vibrations: The Physics of Music
You Should Write a Book
Write a Book that Will Sell
Feel Great Feel Alive
Science 101: Physics (Smithsonian Inst.)
The Fantastic Physics of the James Bond Movies
Baby's in the Doghouse
The Issac Newton School of Driving: Physics and Your Car
Quantum Legacy: The Discovery that Changed our Universe
Alien Life
Cosmic Time Travel
The Mystery of Gravity
Creation: The Story of the Origin and Evolution of the Universe
Stairway to the Stars: The Story of the World's Largest Observatory
Search for a Supertheory
Invisible Matter and the Fate of the Universe
The Vindication of the Big Bang
Colliding Galaxies
Chaos in the Cosmos
Concepts of the Universe

Young Einstein

A Novel of Einstein's Early Life

by

Barry Parker

Drawing by Lori Beer and Laraine Hatch

ISBN 978-0-9764076-4-5

Library of Congress Control Number:
2009912771

Acknowledgements: Title page by Lori Beer

Background Photograph courtesy of NASA

Visit the Author's Website at
www.BarryParkerbooks.com

His email is parkbarr@isu.edu

Chapter 1

1889 - 1894

On a fall day in 1889 nine-year old Albert Einstein walked slowly home from school. Dressed in a dark jacket that extended almost to his knees, he kicked occasionally at the cobblestone street as his dark curly hair blew in the wind. Dark clouds swirled overhead as the setting sun broke through the scattered silvery mist near the horizon creating long golden rays that struck the street in front of him. Smoke rose from the chimneys around him, arcing sharply as it was caught in the wind. He passed a shopkeeper and smiled. From all appearances he was cheerful and easygoing, but to those who knew him well he was strong-minded and stubborn, and most of them would say he was different – always asking questions, frequently staring off into space. This difference stemmed mostly from an intense natural curiosity. Where most boys accepted the world around them, Einstein questioned every aspect of it. Why did things fall to the ground? Why was the Earth round? Why did the sun and moon appear to move across the sky? Why did a compass needle always point in the same direction? Coupled with this was a passion for puzzles and an overriding enjoyment in building complex and intricate structures with blocks, cards and sticks.

Several boys ran past him playing a game, but he ignored them.

"Do you want to play?" one of them yelled at him.

Einstein shook his head.

"You're a bore. You never want to do anything."

"Go to hell," Einstein yelled at him as his dark eyes narrowed in annoyance.

Finally he could see his house in the distance; it was much larger than the house his family had lived in when they were near the center of Munich, and Einstein liked it much better. He now had his own room and a large yard to play in, and best of all: Uncle Jakob was close – next door in an adjoining house. Jakob was an engineer, and Einstein could talk to him in a way he couldn't talk to his father. His father didn't understand mathematics and science the way Jakob did, and although Einstein felt close to his father, his interest in science and mathematics had drawn him increasingly to his uncle.

As he approached his house he looked past it to the back yard. Several large trees stood in it. He had occasionally climbed one of them, but wasn't as adventurous as some of the neighbor boys who liked to see who could climb the highest. He preferred to watch. Around the edge of the yard were flowerbeds that earlier had been filled with a dazzling assortment of violets, azaleas and wildflowers, but with the cold nights all that remained were tangled brown masses.

His father's business was only a short distance away on Lindwurmstrasse – a large, impressive building in which dynamos, armatures, arc lamps and other electrical equipment was manufactured. Uncle Jakob was the technical advisor, and inventor; Hermann, Einstein's father, looked after the business and sales.

As Einstein went to go into the yard he saw Jakob approaching from the direction of the Einstein brothers business. He was walking fast with a military-like stride, a habit he had picked up while serving in the Franco-Prussian war several years earlier. He was tall and well-built with short dark hair that was starting to recede at the temples. Beneath his sharp delicate nose was a finely trimmed mustache. Although he smiled and laughed frequently, he had a serious side and tended to worry unnecessarily. "Albert," he called out to him.

Einstein stopped and waited for him. "Why aren't you working?" he asked as Jakob approached. He was surprised that Jakob was in shirtsleeves despite the cold weather.

"I left some plans at home. I was going to get them. Come on over, I 'll show them to you."

Einstein followed him into the kitchen of his house. It was large with a cast-iron cooking stove near one end; dark chair rail lined the walls with heavily varnished brown boards set vertically below and tan wallpaper above. Jakob's wife was busy over the stove as they walked in; she greeted them. "Let me get your coat," she said rushing over to Einstein She was small, and as frail as a bird, but she had a sweet disposition. To Einstein she was always busy – slicing sausages, stirring soup, kneading bread, and when not cooking she was sewing or knitting.

Jakob took the plans down from one of the high cupboards that lined two walls of the room. He spread them out on the table.

"What is it?" Einstein asked after staring at the complex drawing for several seconds.

"It's my latest invention ... a new dynamometer," Jakob said proudly. His eyes shone as he scratched the back of his head.

Einstein was impressed. He leaned forward so he could see it better.

"I've applied for a patent," said Jakob. "I'd like to start

LARAINE HATCH

manufacturing it. It's much better than anything on the market."

"DC or AC?" asked Einstein.

"It's DC, but as far as I'm concerned the future of electricity is in DC. AC is just a passing fad."

Einstein knew enough about electricity to understand some of the features. Jakob filled him in on the rest, explaining how the dynamo worked, and how it was better than the ones already on the market. When he finished he asked, "Did you solve that problem I gave you?"

Einstein smiled. "It was easy." He looked around for a pencil and paper. Jakob jumped up and got them for him. Einstein wrote down an equation and quickly solved it.

Jakob shook his head in admiration. "Pretty good for a nine-year old. You'll become a professor yet." He folded up the plans, then as an afterthought he asked. "How are you liking your new school?"

Einstein had started at the Luitpold Gymnasium a year earlier. The students were predominantly Catholic; he was one of the few Jews, and although he was a loner and felt he didn't fit in, he had experienced little racial discrimination. In addition to the usual classes in Greek, Latin, German, French, mathematics, geography, literature and science, he had been taking special tutoring in Judaism.

Einstein thought about Jakob's question. He shrugged. "I don't like it much."

"What's wrong?"

"There's too much Greek and Latin and not enough mathematics and science, and I don't like some of my teachers." He paused. "I like my mathematics teacher and my literature teacher, but the others are too strict."

Jakob smiled. "Discipline is important. That's why Germany is such a great country, and has such a great army."

"I don't care. I don't like it, and I don't like my athletics class. I get dizzy when they make me exercise too much."

Jakob looked surprised, then chuckled. "But you love hiking in the mountains."

"That's different."

Jakob put his hand on Einstein's head. "Don't worry. You'll do all right. You've got a good head on you." He stood up. "I've got another problem for you."

Einstein looked up at him in anticipation. He was pleased.

"This one is a little harder than the last. Do you want to copy it down?"

"Okay."

"Find a fraction such that twice the denominator is greater than the square of the numerator by 2, and the product of the sum and difference of the numerator and denominator is 325." He paused. "Have you got it?"

Einstein nodded.

"I'll give you until tomorrow to solve it."

"I'll do it right now," said Einstein cockily as he began writing down an equation. Jakob watched as he solved the problem. In less than two minutes he had the answer. "6/19's" he yelled jubilantly.

"By gully ... you're right."

Einstein jumped up and down with joy. "I told you I could do it."

Albert, come down here," Pauline Einstein yelled up the stairs. She was a tall, strong-minded woman with a broad nose, gray eyes, a full figure and a passion for music. Einstein loved to listen to her play the piano. He particularly like Schubert and Beethoven.

When Einstein got to the bottom of the bottom of the steps, both his mother and his sister, Maja, were waiting for him.

"We're having a guest for lunch," Pauline announced as she pushed back several strands of her unruly hair from her face. "I want you to make him feel comfortable."

"Who?" Maja asked. She was two years younger than Albert and resembled him in many ways, with the same dark curly hair and some broad lips. Her face was round, her skin soft and white and she usually had a twinkle of mischief in her large brown eyes. Like many others she had suffered from Einstein's temper tantrums when he was young. On one occasion he had thrown a bowl at her head, on another he had struck her with a small pickax, digging it into her head, but they were now close. Einstein loved and adored her, and she had strong feelings for him.

"His name is Max Talmud," said Pauline. "He's a medical student at the university. He's Jewish and has little money. It's a tradition in the Jewish faith that we help one of our own whenever we can. This is our way of doing that."

A look of confusion came over Einstein's face. "Why do we do it?" he asked. "We don't follow the other teachings of Judaism. We don't go to Synagogue. We eat pork."

"Albert, you're not supposed to say things like that. You know how your father feels."

"Why does he feel that way? He's Jewish and he should be proud to be a Jew."

Pauline's eyes narrowed in annoyance. "He is. Just because he doesn't go to Synagogue and doesn't pray every day doesn't mean he's not proud of being a Jew."

Einstein stared at her in silence as his lower lip puckered in a pout. His Judaism teacher had told him how important it was to follow the rules of Judaism.

"Get ready for lunch," said Pauline, trying to make sure nothing more was said on the subject. "Your father will be here in a few minutes and Mr. Talmud will be with him,"

She barely got the words out of her mouth when the door opened. Hermann walked in, followed by Talmud. Hermann had a dark mustache that extended beyond his upper lip, a broad, firm jaw and a slightly receding hair line. His hair was cropped short and he wore a monocle, giving him the stern appearance of a Prussian army officer, but in reality he was kind and friendly and loved by all. Talmud was slightly shorter than Hermann. He had dark hair, wore glasses and had the curious habit of twisting the lobe of his left ear when he was thinking. His clothes were worn, and there was a patch on the sleeve of his jacket. But he was a university student, and that excited Einstein.

They had just started the meal when Pauline said with pride, "Albert's number one in his class."

Einstein sighed in embarrassment. "I'm only number one in my mathematics class"

Talmud lowered his fork and picked up a roll. "That's very good," he said as he began buttering it.

"But you're number two in science, and very good in all your other classes," said Pauline.

"Not all of them, Mamma."

"You like mathematics and science?" Talmud asked as he took a bite from his roll.

"I like them the most," replied Einstein.

"Have you taken algebra yet?"

"No, but my Uncle Jakob has taught me. He gives me problems to solve."

As the dinner continued Einstein grew increasingly impressed with Talmud. He was delighted to find that, although he was a medical student, he was interested in both mathematics and science.

"Albert also plays the violin beautifully," Pauline said as the meal was ending. "You'll have to listen to him play. We play duets." Einstein had been taking violin lessons for three years and although he enjoyed music, he hated practising.

"I'll do that," said Talmud.

After the meal Talmud had some time before going back to classes so he walked over and sat down beside Einstein. "What sort of science are you interested in?" he asked as he leaned back in the sofa.

Einstein thought for several seconds. "Atoms and forces. Electricity and magnetism. My Uncle Jakob is very interested in electricity; he's an engineer." Einstein paused. "He has invented several things. His inventions are interesting but I'm more interested in the nature of forces and atoms. When I was very young Papa bought me a compass. It seemed strange to me that the needle always pointed in the same direction. I know now that it is due to the Earth's magnetic field, but then it seemed like magic, and I wanted to know why it acted so strange. And I still want to know deeper truths." He stopped abruptly and looked directly at Talmud. "Do you believe there are atoms?"

Talmud smiled "You can't see them ... but yes, I believe they exist."

"Our teacher told us that not all scientists believe they exist."

"I've heard that too," said Talmud, nodding.

"I would like to prove they exist," said Einstein.

Surprised by Einstein's intensity, Talmud chuckled to himself, wondering if he should take him seriously, but when he looked at his face closely and saw the determination in it, he knew he was serious. "How would you do that?"

he asked.

Einstein shrugged. He had given the question considerable thought, and had some ideas, but was reluctant to discuss them. "I'm not sure yet."

Several days after Talmud's first visit Einstein knocked on the door to his father's study.

"Come in," said Hermann.

Einstein stepped in. "Papa, I would like to speak to you."

Hermann was seated at a large mahogany desk at the opposite end of the room. A gas lamp was on above him, spreading soft yellow light over the desk. He pushed his chair back and smiled affectionately at Einstein. "Come over here son, what's on your mind?"

Einstein walked up to him. "Do you believe in God?" he asked.

Hermann was taken back by the question and was unsure how to answer. He respected his Jewish origins, but he had little use for the customs and rituals that came along with the Jewish religion. He rarely, if ever, read the Bible, even though he was familiar with some of the stories in it from his youth. "Of course," he said finally.

"Do you believe in the God of the Jews?" Einstein said as he stared intently at his father waiting for the answer.

"Why are you asking me these questions?" Hermann demanded. "Do you think I am an atheist?"

"You don't go to Synagogue. You don't pray. You eat pork"

"Albert ... Albert," his father said good-naturedly. "You shouldn't worry about such things."

"My Judaism teacher said that all good Jews go to Synagogue, and they pray, and they don't eat pork." He

paused. "Are we good Jews?"

"Of course, of course," said Hermann. He put his hand on Einstein's shoulder. "You're too young to worry about such things."

"But you send me to learn about Judaism. Why, if you don't believe in it?"

"It's a state law."

"Is that the only reason?"

Hermann was beginning to feel uncomfortable. He knew there was little he could say that would be of comfort to the boy. He didn't have strong feelings about religion, but he did feel a sense of duty in relation to his children. He looked up at Einstein from his chair and smiled. He knew he had failed him in many ways; he couldn't answer his questions about science and his business left little time for family activities, but he felt a strong affection for him. "Son," he said, "if you believe something, and are convinced it is true, you should stick to your beliefs. Don't let anyone tell you different. You've only got one life to live ... and God only knows ... it's mighty short. You've got to live as it's your nature to live, and you've got to believe as it's your nature to believe." He squeezed Einstein's shoulder gently. He was silent for several seconds, then started to continue, "I'm not sure--"

Einstein interrupted him. "I believe that God is the king of the Jews, and that he created the world, and that he is a wonderful God."

"And you should," said Hermann.

Realizing that his father was uncomfortable, Einstein said, "Thank you for talking to me Papa." He turned and walked toward the door; pausing before it he said "I'm sorry Papa, but I had to know."

Talmud continued coming for lunch every Thursday. Over time he brought Einstein several books. There were popular science books by Aaron Bernstein and Ludwig Buchner. Jakob also gave him several mathematics books, and Einstein studied each of them carefully, working all the problems at the back of the chapters.

He continued discussing science with Talmud, and as he learned more and more a change come over him. His religious beliefs began to waver. One day when he was in the middle of a scientific discussion with Talmud he suddenly asked him, "What do you think of the Bible?"

Talmud straightened up and took a deep breath. "It's a great book ... the word of God. Why do you ask?"

Einstein looked troubled. "Why are so many things in it in conflict with science?"

"What are you referring to?"

"The story of Genesis," said Einstein. "In it the world was created in seven days. Scientists believe it took much longer than that. Some believe it may have taken thousands of years ... maybe millions. And God separated the darkness from the light on the first day but he didn't create the sun until the fourth day. And when Cain was thrown out of the Garden of Eden, he went off and got married. Where did this other woman come from? I thought Adam and Eve were the first people on Earth. And there's the story of Jonah and the whale. Do you believe Jonah would have stayed alive in the belly of a whale for three days and three nights?"

"I believe in an almighty God," said Talmud. "He can do anything."

"He created the world ... the universe?"

"Yes, He did."

"So, why are there so many problems? Why am I so confused?"

"These are things you have to deal with. You have to have faith."

"But I have faith, and I am still bothered."

Talmud sat back in his chair. "If you have true faith, you wouldn't be bothered."

"Papa doesn't have faith."

"He believes in his own way."

Einstein sighed. He didn't like the answers, and he knew that his feelings were going to have a serious affect on his life. He was preparing to become a bar mitzvah, which would make him a full member of the Jewish community. But with his doubts, he could no longer go through with it and he knew that several people would be disappointed.

In the fall of 1891 Einstein started his third year at the Luitpold Gymnasium. He sat back in his seat waiting as the texts for the year were being handed out. Several students had been designated to distribute them.

Einstein watched as they piled up on his desk. The only ones he was really interested in were the mathematics and science books. He shoved the others inside his desk and began thumbing through one of those remaining on his desk. It was a geometry book. He had never taken geometry and was curious about it. He looked over the table of contents, then began paging through it. Diagrams covered many of the pages.

Finally he turned to the beginning and began to read. Several axioms were presented, then used to prove a theorem. Einstein was intrigued with the elegance of the proof. He read several more pages and tried some of the exercises. He was pleased to find he could solve them.

As he put the book down he smiled; he would show it to Talmud when he came on Thursday. He continued reading and working exercises as he waited. When Thursday finally came he rushed Talmud up to his room.

"I have something to show you," he said. He grabbed the geometry book from the small desk in his room and held it up for Talmud to see. "It's a geometry book. Take a look at it," he said excitedly.

Talmud took it and began thumbing through it. "How much of it have you read?"

"Only the first twenty pages or so. I won't be studying it in class for several months, but I couldn't wait. I've worked some of the exercises. Here ... I'll show you." He reached down to the top of the desk and took several pages from it, then pointed out which exercises they were in the book.

Talmud looked them over, and seeing that they were correct he smiled, "Excellent," he said.

"I'm sure you could use the same sort of proof in science," said Einstein. "I could use proofs like that to prove that atoms exist."

Talmud chuckled skeptically. "I think that would require a different kind of proof ... an experiment."

"That may be so, but there may be things around now ... like the axioms in the book, that I could use for the proof."

"It's possible," said Talmud. "But I think it will be shown by experiment.

"Experiment and theory go hand-in-hand," said Einstein emphatically. "An experiment may show something that seems completely unrelated to the structure of atoms, then somebody sees a relationship. They put two and two together and show that because of this experiment atoms have to exist."

"But it takes years of study before a person is in a position to do this. You have to have a thorough understanding of nature and science; you have to understand all the experiments that have been done, and what their relevance is. You are still very young."

"I'm not too young to be thinking about it, and to be preparing myself. A person has to have a willingness. You have to do a lot of thinking about these things. Only through deep thinking will they be revealed, and I believe I am capable of doing this deep thinking."

"You're only twelve years old, Albert."

"Twelve years old isn't too young to begin preparing yourself."

"That's true and you've done a lot already and I think you should continue." He put the sheets down on the desk. "I've got another book I would like to bring you."

Over the next several weeks Einstein worked every exercise in the geometry book and continued studying more advanced books on trigonometry and analytic geometry.

Einstein put his violin down as his mother entered the room. He had been practising for over an hour, and during much of the time he had been brooding. His music teacher had criticized his playing during his last lesson and put him in a bad mood. He had been taking lessons since he was six, and had never been thoroughly satisfied with any of his teachers, and although his present teacher was better than most, he was also a stickler for perfection.

"Mamma," he said. "I'm going to stop taking lessons."

His mother looked at him in shock. "Albert, you can't. You're playing beautifully now. I thought you were enjoying Mozart. Just yesterday you told me how much his music affected you."

"That's true, I love Mozart, but I hate the etudes, the exercises and scales my teacher gives me, and the way he keeps forcing me to play a piece over and over. He is

destroying my love for music."

"But Albert, you have to play a piece many times. You have to practise it over and over until it is perfect. Only then is it truly yours."

"I don't care if it is perfect. I'm not going to be a concertmaster." He picked up the sheet music he had been playing and looked at it briefly, then put it back on the rack. "The lessons are expensive. I'll be saving you money."

"The money is not important. What is important is that you learn to play the violin well. You've improved a lot. I don't know why you want to quit." She paused as a look of determination came over her face. Why was he so stubborn? Maja wasn't that stubborn, and he certainly didn't get it from his father; for all his other faults, he wasn't the least bit stubborn. "I forbid it," she said looking him fiercely in the eye.

"No, Mamma, I'm going to quit. You can't force me."

The remark caught her by surprise. He had never defied her before, and she wasn't quite sure what to do. She stared at him for several seconds. "Your father will not be happy."

"Father has no great love for music. I think he will be indifferent." Einstein looked down at the floor, then back at his mother. "It's not that I'm going to quit playing. I will always play. I may even play more. I love Mozart, Schubert and Beethoven. I won't stop playing them, and I think away from my teacher I will play them more beautifully, with much more feeling."

"You promise to practise every day?"

"No."

Pauline's eyes widened. "You said you would."

"I will not be practising. I will be playing, and I will play as much as possible, but it may not be every day."

By the early 1890's Hermann and Jakob's business had become relatively successful. Electricity was now coming into its own, and it was an exciting time in the electrical industry. Whole towns were being lit for the first time by electrical lights. As the only electrical manufacturer in Munich, Hermann and Jakob expected to get the contract to light the city. But there was a problem: their experience was mainly in DC, and when it was shown that long distance transmission of electricity via AC was much less expensive than DC, they knew they had to make some changes if they were to compete with the large out-of-town companies, and that required considerable capitol. So in 1893 they borrowed against their homes and any other collateral they could come up with.

Jakob sat on a stool in the office with his hand under his chin. "When will we find out who gets the contract?"

"The notice should come in today's mail." said Hermann.

Jakob took out his pocket watch and looked at it. He tapped his fingers on the desk. "What do we do if we don't get it?"

"You worry too much. We'll get it. We're the only electrical contractors in Munich big enough to do it, and we've set up electrical plants in both Varese and Susa."

"But they were both DC – three conductor plants. Munich wants AC."

"We've made the changes. We won't have any problems with AC."

Both men turned as they heard a noise near the door. "It's the mail," said Hermann.

Jakob rushed to the door. He quickly flipped through the mail, stopping suddenly on an envelope with the letterhead

"City of Munich."

"It's here," he said, taking the envelope in his right hand and putting the others down. "Do you want to open it?"

Hermann took the envelope from him, stared at it for several seconds, then tore it open and began to read. His face turned white. "The contract was awarded to Schuckert of Nuremberg."

"My God," said Jakob. The two men stared at one another for several seconds without saying anything, then Jakob broke the silence. "We won't be able to pay off the loans."

Hermann shook his head slowly. "We'll lose our houses. How can I tell Pauline and the children?"

Hermann and Jakob discussed the problem over the next few days. They finally decided to liquidate everything in Munich and start a new plant in Pavia, near Milan, Italy. The two towns they had supplied electrical lighting for were both in Northern Italy, near Pavia, and Garrone, their Italian representative, had encouraged them to move to Italy. He would put up some of the capitol needed for their new venture. One of Pauline's relatives also agreed to put up some money.

The evening after Hermann and Jakob reached their decision Pauline made the announcement at the dinner table. "Albert, Maja," she said. "I have some bad news." Albert and Maja waited in anticipation as Pauline paused. Finally she continued. "Your father has had some business problems and we will have to sell everything, including the house. We will be moving."

"Where to?" Maja asked with widened eyes.

"Milan, Italy," said Pauline.

Einstein was surprised, but inwardly he was pleased. He

was glad to be moving. School was becoming an increasing burden to him, and he was beginning to hate it. A move would be a great relief.

"Maja, you will be coming with us," Pauline continued. She turned to Einstein. "We have decided you will stay, Albert. You are close to receiving your Gymnasium diploma, and you must finish and graduate."

Einstein looked at her in alarm. He couldn't believe what he was hearing. "I don't want to stay. I can finish my schooling in Milan." He started to choke as the words came out.

Hermann intervened. "Albert, we know how you feel, but it is best that you stay here."

"I hate the schools here," Einstein yelled. The anguish in his face was echoed in his voice. "I hate them. I don't want to stay." He pounded the top of the table with is fist.

"We've already made arrangements to board you out," said Pauline, eying him nervously, but determined not to back down.

"I want to go with you," cried Einstein stubbornly.

For several seconds no one said anything and Einstein saw a slight glimmer of hope. He looked from his mother to his father, knowing that he was is only hope. If his mother's mind was made up she would not change it. His father looked at him sympathetically, then looked down at the table. Einstein knew it was hopeless to argue any further.

"You also have a military obligation when you graduate," said Pauline. "You'll have to attend to it, and you can't leave Germany until it is fulfilled. It is forbidden."

"I hate the military." He closed his eyes as he let his breath out slowly.

The following Thursday Einstein rushed down the street

to meet Talmud as he approached. He was out of breath when he reached him. "Max," he said. "We're moving."

A frown came over Talmud's face. "Where to?"

"Everyone is going to Milan except me. Papa's business has failed." Einstein told him he would be left behind at a boarding house so he could finish his schooling. "I won't be seeing you anymore." His eyes glassed over. "I won't be able to talk to you anymore. Our discussions ... I don't want them to end."

Talmud shook his head. "You've learned so much calculus, and so much science I can hardly follow you anymore. You're so far beyond what I know now, I don't think I've been of much help lately."

"Yes, you have," said Einstein emphatically. "You've challenged me. You've helped me see things much more clearly. When you make comments I think about them. I consider all the possibilities. I ask myself again and again if my views are really right. Too many scientists look at the world through blinders. I don't want to be like that. You've helped me more than you think."

Talmud was grateful for the compliment, but he knew that Einstein would have persevered without him, and would continue to do so after he left. "I'll be graduating very soon anyway," he said, then after some hesitation, "I hope we'll always be friends, and I know you have a great future ahead of you ... you can do great things if you put your mind to it. You have what it takes."

Over the next few weeks Einstein watched as everything at the Einstein factory was sold off. Then the buildings were auctioned, and finally the house that he had lived in and loved so much was sold.

Box after box was taken to the train station, then finally

the day of his family's departure came. They were all assembled on the platform at the station. Einstein could hear the locomotive approaching in the distance. Across the haze of early morning he watched with numbed lips and a hollowness in his stomach. Visions of his life over the past few years floated through his head: Maja's happy, smiling face, the large rooms and garden of their house, the trees, his mother's music and his father's soft laughter. He would miss all of them desperately. But he couldn't break down; he was 15 and determined not to show the emotion he felt in his heart.

The locomotive drew into the station, looming large above them. Steam hissed from near its wheels, then he heard the grinding of steel on steel. Finally it came to rest. Einstein looked over at his mother's white, sullen face; she was watching the train intently, and he knew that she had to be as heartsick as he was. She had loved their life in Munich -- particularly the large house.

Maja looked over at Einstein and smiled, then quickly lowered her eyes to the platform. He had always felt close to her, and he felt even closer now. Several passengers got off the train. Hermann and Pauline waited until they were clear, then Pauline turned to Einstein. "Albert, I want you to remain number one at school. Your aunt will be checking on you, and we will write to you. Be sure you write." She hugged him briefly, but there was little show of affection.

His father walked up to him. "Good-bye Albert." He tried to say more but couldn't find the words. Disliking any show of affection in public, he held out his hand. As Einstein took it he looked up at his father's face. Words weren't needed; his father's eyes and smile said everything. He shook his hand and smiled back.

Maja had tears in her eyes as she turned to him. "Oh

Albert, I'll miss you so much," she said as the tears began to flow down her cheeks.

"And I'll miss you," he said, straining to hold back his emotions.

He watched with a heavy heart as they boarded the train. He had never been separated from his family before. He had never felt such loneliness, such desolation. The world seemed empty, forlorn, forsaken and he felt abandoned. The landscape shimmered as he watched the train grow smaller in the distance. He bit his lip, almost drawing blood as he rubbed his throbbing temples. Finally he turned and began to walk away.

Chapter 2

1894 - 1895

Einstein walked slowly back to his boarding house, kicking at the ground every so often in disgust, ignoring the people passing him on the sidewalk as he silently brooded. The heavy dark clouds overhead added to his gloom. As he walked a light drizzle began to fall but he never felt nor saw it. The empty feeling in his stomach had not gone away, and his chest still ached. His life had been turned upside-down and he resented it.

Had he been betrayed by his parents? Was their desire for him to succeed greater than their love for him? He ran his hands through his thick black hair as he thought about them. His mother's parting words were still ringing in his ears. "I want you to remain number one in your class." How could he remain number one when he hated school so much, and had over a year left to go? And beyond that, because of his military obligation, he would be able to do little more than visit them.

The house he was boarding in was dreary compared to his parent's house; the exterior was stone and gray brick, and the interior was like a mausoleum – damp, dingy – and there was a strange odor throughout it. Furthermore, his landlady was a old as his grandmother, but so different. She had missing teeth, and a large mole on her face and although she was polite to him, she seemed cold and distant.

"Hello Albert," she said stiffly as he walked in the front door.

He tried to smile but couldn't, and in his state of mind it was difficult to speak, so he merely nodded as he passed her on the way to his to his room which was on the second floor of the house. He climbed the stairs two at a time, and closed the door as he went in. The room was small, about half the size of his former room, with a narrow bed along one wall and an old desk against the opposite wall. The wallpaper was old and yellowed, and was peeling off the walls in places above the baseboard.

Einstein sat down at the desk. The window above it was open and he could hear noises from the street below. A carriage was passing, and in the distance he heard the whistle of a train. He had managed to bring most of his science and mathematics books with him and they were in a small bookcase beside the desk. Earlier he had read Kant's "History of Nature and Theory of the Heavens" and it had intrigued him. He was particularly impressed by two ideas that Kant had put forward. The first was the suggestion that our solar system formed from a giant nebular disk, the second was that our galaxy, the Milky Way, was an "island universe" of stars, surrounded by millions of similar island universes. Talmud had brought him another of Kant's books titled "Critique of Pure Reason." Einstein pulled it out of the bookcase and began reading. After a few pages, however, he leaned back and closed the book. It was hard to concentrate when he was so distracted. Too many things had happened and he couldn't get them out of his mind.

With his family gone life seemed so empty and meaningless. He opened the book again and forced himself to read. He couldn't let his mood distract him. There was so much to understand and so little time.

It was a cold, dull November morning with snowflakes floating down like feathers out of a slate grey sky. Einstein grabbed his heaviest coat and a cap. He didn't like wearing a cap, but the dark, ominous clouds warned of heavier snow later in the day. Several weeks had passed since his parents had left and things had got no better at school. They had written several letters, and although eagerly awaited, they had left him remorseful and melancholy. His mother had written about their new house, their new life, and the beautiful countryside around Milan. Maja had also enclosed letters and they made him particularly homesick.

As he approached the school yard one of his classmates ran by him and grabbed his cap. Turning, he saw Heinrich Schweiller headed for the large stone gate of the school yard with the cap in his hand. He cursed as he gave chase; Schweiller had given him trouble before and Einstein didn't care for him. Just inside the gate Einstein caught up with him and grabbed the back of his coat. Schweiller stopped, but held the cap out of reach.

"Give it to me," Einstein yelled at him.

"Do you want to fight for it?" Schweiller yelled as a challenge. He was slightly taller than Einstein, but thin and had a permanent scowl on his face.

"No," said Einstein in irritation.

"Why not? Are you afraid?"

Einstein glared at him. "No, I don't want to fight. Just give me the cap."

Schweiller held the cap in his left hand while he fended off Einstein with his right. A crowd started to gather around the two boys. Einstein finally tired of yelling at him; he stood glaring at him, saying nothing as the crowd got larger. He could feel his temples throbbing, and would like to have hit

LARAINE
HATCH

him. He took a deep breath as he tried to calm himself; he didn't want to fight, even though he was sure he would get the better of him. Fighting was not in his nature.

Suddenly somebody grabbed the cap out of Schweiller's hand. Turning, Schweiller saw who it was and decided not to challenge him. Taking a last look at Einstein he made a face and ran off.

"Thanks Hans," said Einstein as the boy handed the cap to him. Hans Reibacher was one of the few people Einstein talked to at school. He was blond, had short hair and was taller than most boys his age.

"I hear your parents left," said Hans.

Einstein put his cap back on his head as he took another deep breath. He sat down on a nearby step; Hans sat down beside him. "Yeah," he said, "but I'll be joining them when I graduate."

Hans frowned. "That's a long ways off."

"Not that far."

"What about military service?"

Einstein stared at him briefly, then shook his head. "I'm not going to serve."

Hans looked at him in surprise. "How can you get out of it?"

"I thought it over," said Einstein. "I've decided to renounce my German citizenship."

Hans's mouth dropped. "You wouldn't."

"I would. Why do I want to remain German. I'm not a German at heart, I'm a Jew. And I don't want to serve."

"You can't renounce your citizenship without your parent's permission. Have you told them?"

Einstein shook his head. "No, but I will."

"Will they sign for you?"

"When I explain, I'm sure they will."

Einstein ran through the front door. Whenever there was mail for him, his landlady left it on a small table near the door. An envelope sat on the table. He tore it open. Two letters were enclosed; one from his mother and one from Maja. He began reading his mother's letter. They were planning on leaving the house they had bought in Milan and would be moving to Pavia in the spring so they could be closer to the factory. Einstein wondered if that was a bad omen. Was their new business venture going well? He read the letter through quickly to see if there was any mention of the business, but his mother said nothing. She appeared to be unhappy about the move but said it would be more convenient for his father. Pavia was about 25 miles from Milan, and traveling had become a problem. Einstein wondered if she was hiding their problems from him, as he was hiding his feeling from her. He had never said anything in his letters about the way he felt.

When he finished reading his mother's letter he opened the letter from Maja. She was always so innocently joyful and enthusiastic; reading her letter brought a tear to his eye. She made several references to their home in Munich, and how much she had enjoyed it. As Einstein put her letter down he thought about their house; it wasn't far away, but he hadn't visited it since they left. He was sure it would have been too painful, but suddenly he had a yearning to see it. He would go after school.

It was late in the day when he approached the street he had lived on. In the distance he saw smoke rising to the leaden sky above; it was a much heavier stream than those rising from the numerous chimneys around him. Somebody had to be burning something in the open. Wondering where it was coming from, he increased his pace. He passed familiar landmarks – the view of the Synagogue in the distance, the large grey house next door – then he saw an empty lot with a fire near the other end and a feeling of shock came over him.

He rubbed his eyes. The house they had lived in was gone; even the trees were gone. Everything had been leveled, and in their place were the foundations of several buildings.

Einstein stood with his mouth open. Several people were working on the foundations. He ran toward one of them. "Where is the house that was here?"

The worker, who was bending down, looked up at him in confusion. He had a tooth missing in the front.

Einstein repeated the question as the worker stared at him. Finally he replied, "Cleared away. New apartments are going in."

Einstein stared at him in disbelief. "Why would they destroy such a beautiful house?"

The worker shook his head and grunted, then without saying anything he went back to work.

Too stunned to say anything further, Einstein turned and walked away. His home – the only home he had ever known – was gone. He thought of his room with the large window that overlooked the back yard, the beautiful dining room with his mother's piano in it, the large kitchen. All were gone. It was like a bad dream.

Einstein's relations with his teachers continued to deteriorate. He could barely tolerate most of them, but he liked his mathematics teacher and he respected his literature teacher. His mathematics teacher had helped him with some of the exercises he had worked outside of class, and his literature teacher, unlike most of the other teachers, had a liberal mind and was open to questions.

Einstein approached his mathematics teacher as the class let out. "Mr. Zametzer, I was wondering if you would help me with a problem."

Zametzer smiled, "I'm sure it's not one of the problems from class. What are you working on now?"

Einstein held up an introductory calculus text.

"You're studying such advanced material I can barely follow you."

Einstein read a problem on transcendental functions to him out of the book. "There are two maxima," he said. "Which one do they want?"

Zametzer rubbed his chin as he looked at the problem. With some difficulty he worked the problem for himself and explained it to Einstein.

"You're doing extremely well in your mathematics class, Albert" Zametzer said. Then hesitatingly he asked, "Are you having any problems in any of your other classes?"

Einstein shook his head slowly from side to side, but he knew he was lying.

"Several of the teachers are talking about you," said Zametzer, as if reluctant to tell him.

"What are they saying?" Einstein asked

"They're saying you're a troublemaker."

Einstein looked down at the floor. His face became like marble and his brown eyes narrowed as he clenched his teeth. So they thought he was a troublemaker. It didn't surprise him, but he knew he had caused no trouble. He thought about each teacher in turn, lingering on the ones he disliked the most. They were the ones causing the trouble, and there was no reason for it. He had never been disrespectful and had never talked back to them.

Zametzer saw the look on his face and was sympathetic,

but there was little he could do. He put his hand on Einstein's shoulder. Einstein looked up at him and smiled briefly. "Thanks for helping me," he said.

As he walked down the cobblestone street away from the school he thought again about what Zametzer had told him, and his feeling of hatred for the school intensified. How long could he tolerate it? He had to get away. But how? He couldn't leave without an excuse; he might be picked up for truancy. Furthermore, he eventually had to complete his schooling and get a diploma, but he could do it in Milan, or Pavia.

He mulled over the excuses he could use, but in the end it seemed that there was only one that would be useful to him. The only one anyone would believe would be an excuse for reasons of health. He would have to visit his doctor.

Einstein sat in the waiting room with two other patients. He had made the appointment several days earlier. Furnishings in the room were sparse, and the walls were barren except for a small picture, a landscape. Three chairs stood next to the wall on the side he was on and three against the wall opposite him. Four were occupied. Between them was a coffee table that stood on a well-worn rug. Dr. Talmud was Max's older brother, and he had been the family doctor for many years. Einstein had been to him twice for minor ailments, and had found him friendly and understanding. He was sure if he explained things to him, Talmud would help.

Einstein watched as the other patients were called in. Finally the nurse announced his name, and led him to a small room. Moments later Dr. Talmud walked in; he resembled Max, but was shorter and heavier, and much less serious. "Albert, I'm delighted to see you," he said as he crossed the room and sat in a chair. "Max has been telling me so much about you. He says you're a whiz in mathematics and science, and have a great future ahead of you."

"My future does not seem very bright right now," said Einstein nervously.

Talmud smiled, then a serious look came over his face. "It must have been hard on you. I heard. Your family has gone to Milan." Max had told him they had left Albert in Munich to finish gymnasium. He grimaced as he glanced at the chart in his hand. "How can I help you?"

Einstein hesitated. "I'm not sick physically, but I don't feel well. I haven't felt well since my family left."

"That's understandable," said Talmud. "But I'm sure it's mostly emotional. Something like that would be a shock to anyone, but you'll get over it."

"I want to join my family," Einstein said quickly.

Talmud sat up in surprise. "You mean for a vacation?"

Einstein shook his head. "No, I want to join them permanently."

"Have you talked to them about this?"

"No."

Talmud turned sideways in his chair and banged the chart against his knee several times as he studied Einstein's face. "I think that might be impossible under the circumstances."

"Nothing is impossible," Einstein said stubbornly.

Talmud sat looking sympathetically at Einstein. Finally in a low voice he asked, "Is there anything I can do to help?"

This is what Einstein was waiting for. "Write me a note saying I am on the verge of a nervous breakdown," he said, "and that I must go to a more ideal climate, such as that in Northern Italy, or I will have a complete breakdown."

The suggestion took Talmud by surprise. "It's not something I should do, but if I can help I will. Your family was wonderful to Max and I know he enjoyed your company. I would like to pay you all back." He began writing the note, then interrupted himself. "How are you planning on finishing your education? Max said it would be a shame if you didn't go to university."

"I'll find a way."

Talmud finished the note and handed it to him. Einstein thanked him and left. As he walked out the front door he unfolded the note and read it. When he finished he smiled; this is what he needed. Nobody could ignore it; his health was at risk. But he needed another letter -- a letter from his mathematics teacher attesting to his mathematical ability and knowledge. He knew Mr. Zametzer was sympathetic, and there was no doubt that Einstein was far ahead of the class. In fact, he easily had enough mathematics to graduate.

Einstein waited until the class cleared out. Mr. Zametzer sat behind a large desk at the front of the room. He was tall and had only a fringe of hair around his head. Einstein got up and approached him slowly; his heart was pounding but he knew it was something he had to do.

Zametzer looked up as Einstein approached.

"I will be leaving the gymnasium," Einstein announced.

"Why?" Zametzer asked in confusion.

Einstein handed him the note. Zametzer read it though slowly, then raised his eyes. "You look all right to me."

"I don't feel all right."

Zametzer shifted back in his chair. "I know your family has left and it must have been a shock to you, but I find it hard to believe you are on the verge of a nervous breakdown." He leaned forward over the desk. "I hate to see you leave. You are doing extremely well in class, and I'm sure you are doing well in most of your other classes, despite what they are saying about you."

Einstein had no intention of changing his mind. He liked Zametzer and was thankful for the help and encouragement he had gotten from him, and he knew he would miss him, but he had made up his mind. "There's something I would like you to do for me," Einstein said hesitatingly. He cleared his throat, then continued. "I would like a letter stating that I am beyond the mathematics level required for a gymnasium diploma. As you know, I have studied both differential and integral calculus on my own."

Zametzer was reluctant. "How would you use it? Of what good would it be to you?"

"It will be helpful." Einstein replied.

"What are your plans?" Zametzer asked. "Surely you're going to continue with your studies. You should go to university."

"I'm not sure what my plans are right now," said Einstein, "but I do plan on going to university."

Zametzer wrote the note and gave it to him. Einstein was pleased. All he had to do now was present these two letters to the principal.

Much of Einstein's misery centered on his homeroom teacher, who was also his Greek teacher. Not only did Einstein dislike him, but he resented having to take Greek -- sure that he would have little use for it when he graduated. He knew the teacher disliked him, but he didn't care.

Einstein sat at the back of the classroom paying little attention to what was being said. Every so often he would feel the teacher's eyes on him, boring into him with disdain and contempt, but it didn't worry him. He was leaving. When the teacher looked at him he smiled back.

As the class let out the teacher yelled above the shuffle of feet and books, "Albert, I would like to see you at the end of the school day."

Einstein cursed under his breath. He didn't want to see him; he didn't even want to talk to him before he left. Why was he being asked to stay? Even though he wasn't doing well, he wasn't failing. There was no reason for it.

At the end of the day Einstein remained in his seat as everyone else left. His teacher was busy at the desk and never looked up. "What a pompous bugger – so full of self-importance," thought Einstein as he stared at him. After several minutes his teacher had still not looked up, and Einstein thought maybe he had forgotten him. He shuffled his feet nervously. Finally the teacher stood up and began walking toward him down one of the rows between the seats. There was a look of scorn on his face.

Einstein squirmed in his seat. The teacher stood over him for several seconds, glaring down at him, saying nothing. Finally in a stern, military-like voice, he said, "You are a disturbing influence in the class."

Einstein objected. "I have done nothing."

"You sit there smiling like a jack ass. It destroys the respect of the rest of the class for me." He continued staring at him in silence, his eyes boring into him. "You don't need Greek," he said finally.

"I know. It will be of no use to me in later life unless I visit Greece and I don't plan on doing that."

The remark infuriated his teacher. "You don't need it because you'll never amount to anything anyway." He paused. "What are you planning on becoming?"

"I'm not sure," said Einstein,

"It doesn't matter. As I told your father, you will never succeed at anything."

Einstein's eyes narrowed. The remark irritated him, but he said nothing.

"I think it is best that you leave the gymnasium," the teacher said sternly.

Einstein's mouth dropped, and he almost laughed out loud. How ironic, he thought; he was planning on leaving anyway.

"I will leave," he said. He was tempted to add," and I'll be glad to see the end of you and your stupid Greek class," but he kept the words to himself.

The next morning Einstein lay in his bed going over what he would say to the principal. It was early and the house was quiet. He had two notes, and his Greek teacher had asked him to leave. That was all he needed. He dressed and went down for breakfast. His landlady handed him a bowl of cereal as she busied herself at the stove. He was thankful she didn't sit down with him; he didn't like looking at her mole and the hair that grew from it. When he finished he headed for gymnasium. It was overcast and dreary, as it had been for several days. An inch of snow covered the ground in places, but the streets were dry.

Einstein decided to go directly to the principal's office; he would not attend any classes. He had spoken on occasion to the principal, and had found him agreeable, despite his appearance; to Einstein he looked like an army drill sergeant.

Einstein approached the door of his office with trepidation. He knocked.

"Come in," echoed from inside.

Einstein opened the door and entered. The princi-

pal was seated at a large desk.

Einstein handed him the two notes. "I wish to leave gymnasium and join my parents in Italy. The note from Dr. Talmud explains why."

The principal read the note, then looked up at Einstein. "Do your parents know you are coming?"

"No, I have not told them."

The principal frowned as he looked over his glasses. "Won't they be surprised and perhaps disappointed that you have left gymnasium?"

"I suppose they will. That cannot be helped." He stood silent for several seconds, then added, "My homeroom teacher told me I should leave."

The principal grunted as he shifted his weight in the chair. "You are a good student, Albert," he said, nodding his head, "but you are not well-liked amongst some of the teachers."

"I know that."

"Your problem is that you are too cocky, and you give them little respect."

Einstein closed his eyes briefly. "I try to respect them, but in some cases I find it very difficult."

"Under the circumstances it is perhaps advisable to leave the gymnasium, but you should continue your schooling in Italy. You need a gymnasium diploma,"

"Yes."

Einstein sighed in relief as he left the principal's office.

The next morning Einstein packed his belongings and said good-bye to his landlady. He had two bags, one heavy with books, and the other lighter with clothes. He had never taken a long trip such as this before.

He arrived at the station just before lunch. According

to the stationmaster there was a train to Milan later in the afternoon; he would arrive early the next morning. He bought his ticket and sat down on one of the hard benches in the station to wait. As he sat he reflected on his life in Munich. There was little he would miss. His house was gone – replaced by a row of ugly apartment buildings, and Max had left. And the gymnasium – it was a relief to be rid of it.

He dozed off and when he woke the train was pulling into the station. He boarded along with several other people and took a seat near a window. As he stared out the window at the people on the platform he wondered what northern Italy would be like. His mother and Maja had described it in glowing terms. Images of mountains, of hiking and of other outdoors activities went through his head. A feeling of relief overcame him; he was sure anything that was to come would be better than what he had been through over the past few months.

The train pulled out of the station. He stared at the hills, now covered with snow. It was December, and would soon be Christmas; he was glad he would be able to spend Christmas with his family. He contrasted it with the bleak Christmas he would have had at Munich. The snow-covered mountains glistened in the late-day sun as the train chugged slowly up the mountain grades. The train passed rivers snaking their way through snow-covered gulches and ravines, and every so often a waterfall could be seen cascading down the side of a steep mountain. As the sun set behind a distant peak, blue and gray shadows moved across the landscape, creeping like a dense misty fog. It was a beautiful sight, and a new adventure for him, but as he sat staring at it, reality suddenly came back to him, and he began to worry about what his parents would think. What would they say to him? They would be disappointed, but if he assured them he was going to go to university they would understand. One of the most prestigious universities in Europe was nearby – the Polytechnic in Zurich – and he heard that you only needed to pass an entrance exam to be admitted. A gymnasium diploma was not needed. He would take the exam; he was sure he would have no difficulty with mathematics or science, and as for the other subjects ... he would study them over the next few months.

But there was still a problem. His father wanted him to become an engineer, and he was against it; he wanted to teach and do research – theoretical research. It was something he would have to work out. He had learned so much over the past year or so, and he was bubbling over with ideas.

He tried to picture the look on his mother and father's face when he came to the door. They would be confused, but he hoped they would be happy to see him, and Maja would certainly be happy to see him. He looked forward to seeing her.

Finally darkness closed in and he could see little through the window except an occasional light in the distance, and the lights from the few towns they passed through.

The train pulled into Milan early in the morning. Einstein had the address of his parent's house, but he had no idea where it was. He walked into the station and asked the stationmaster, who gave him instructions. It was close enough to walk; he grabbed his two bags and started down the cobblestone road.

Chapter 3

1895

Einstein walked along the narrow streets shifting his bags back and forth between his arms. The one with books in it was much heavier than the other one, and within a short distance the heavier one tired his arm. He passed students on their way to school and sullen-faced workers with lunch pails in their hands on their way to work. Merchants were setting up displays on the outside of their shops. It was chilly, but the rising sun was striking Einstein's back, and it felt warm. A draft horse passed him, pulling a bouncing cart over the cobblestone.

Einstein stopped and rested. According to the station-master's directions his parent's house was just ahead. He could feel his heart beating. He picked up his bags and began walking again. Within minutes he saw the familiar address that he had written to so often. As he approached the house he looked it over carefully. It wasn't as aristocratic on the outside as their house in Munich. The bottom floor was dominated by arches and pillars, and above them were arched windows. Two doors were visible, one somewhat smaller than the other. He wasn't sure which to knock on, but decided on the smaller of the two.

He paused before the door trying to control his beating heart, then he knocked. Several long moments passed without an answer. He could feel sweat forming on his brow despite the cold. He knocked again.

Still no answer.

He began to worry. Maybe no one was at home.

As he went to knock a third time the door suddenly opened. At first all he saw was a large blue apron, then he recognized his mother.

Pauline's eyes opened wide in shock. "Albert, what are you doing here?" she said in a loud voice.

"I left gymnasium."

"You what?" she asked, sure she had heard wrong.

"I left gymnasium."

"What do you mean? Left for good?"

"Yes."

She stared at him for several seconds in disbelief as he started telling her about the letter he had got from Dr. Talmud and how he had taken it to the principal.

"Come in quickly," she said, interrupting him. She gave him a quick hug.

"You're not disappointed are you?" asked Einstein as he followed her through the door.

She turned. "I don't know what to think right now, but I'm glad to see you." She paused as she looked him over. "My ... you've grown since I saw you last." She pulled out a chair for him. "Your father will be disappointed."

Einstein sat down, then looked around. "Where's Maja?"

"She left for school. She won't be home until late this afternoon."

Einstein frowned. He had hoped to see her before she left for school. He looked the room over as he leaned back in his chair; it was the kitchen. His mother had been in the middle of ironing and there was a pile of clothes near an ironing board. The breakfast dishes were still piled in the

LARAINE
HATCH

sink.

"Have you had any breakfast?" Pauline asked.

Einstein shook his head.

She sat him down at the table and got him some cereal. Over the next few hours Einstein told her about his problems at gymnasium and how he had left. He showed her the note from Dr.Talmud and the note from his mathematics teacher.

Late that afternoon Einstein was sitting on the sofa in the living room when his mother ran into the room. "Maja's coming ... quick ... hide," she said excitedly.

Einstein jumped up and ran to the kitchen.

Maja came in through the front door. She was wearing a polkadot dress with a dark jacket over it.

"I have a surprise for you," Pauline said as Maja started to take off her jacket.

She stopped as her eyes widened in anticipation. Then a smile came over her face. "What kind of a surprise?"

"Come here," said Pauline, directing her to the kitchen.

Einstein stepped out.

"Albert," Maja yelled as she ran towards him. She grabbed him and hugged him. "I'm so glad to see you. I've missed you so much." She pulled back and looked up at him. She was 13, but still almost a head shorter than him. "How long will you be staying?" she asked with a look of anticipation.

"I'm here to stay."

Maja screamed in delight. "Oh Albert, that's wonderful."

They talked excitedly for the next hour, each interrupting the other with questions. Pauline could hardly get a word in as they chattered back and forth. Einstein was so absorbed in conversation he forgot about the confrontation he feared the most – that with his father.

Pauline reminded him. "You father will be here soon, Albert," she said. "I don't know what he'll think about you being here." She paused. "There have been problems with the business again. We're moving to Pavia in the spring, as I told you in my letter."

Einstein frowned as a wave of sympathy flooded over him. Life had been so unfair to his father, and he didn't deserve it. He had always worked hard, and despite the setbacks he was always jovial and optimistic. How much more could he take? Einstein knew he couldn't take another setback like the one in Munich, and if the business did fail, what would happen to his mother and Maja?

As he stood in a trance thinking, a noise came from just outside the front door. He scurried quickly to the kitchen. A few moments later the front door opened and Hermann walked in. Pauline and Maja stood staring at him.

"What's going on? Why is everyone looking at me that way?" Hermann said in surprise.

"Are you in a good mood?" Pauline asked.

"I'm always in a good mood."

"No, you're not."

At that moment Einstein stepped into the front room from the kitchen. "Hello, Papa."

Hermann's mouth dropped. "Albert, what are you doing here?"

"Aren't you glad to see me?" Einstein asked as if his appearance was an occasion for celebration.

"Yes, but –" Hermann fumbled for words. "Why aren't you in school?" he asked finally. He was pleased to see him, but confused.

"I dropped out of gymnasium, Papa. I found it very disagreeable after you left."

His father's mouth fell open in surprise. "Dropped out? You can't quit school. You need a diploma. You can't go to university without a diploma."

"I know."

Hermann continued staring at him with a look of mixed affection and confusion. Finally he took off his coat and hung it in the closet, then adjusting his glasses he looked back at him. "What are you planning on doing?" he asked in a questioning voice, still struggling to accept what he had heard.

Einstein shrugged. "I'm not sure. But I have this and I think it might be helpful." He handed the note from his mathematics teacher to him. "And I know I have enough physics to graduate from gymnasium."

Hermann read the note. He looked over at Pauline, then back to Einstein. "What good will this do you? You need a diploma if you are to go to university."

"Not necessarily."

"What do you mean?" Hermann asked in surprise.

Einstein told him he had looked into the requirements of several of the nearby colleges. "The Polytechnic in Zurich is one of the best universities in Europe and it doesn't require a gymnasium diploma. You are only required to pass an entrance exam to attend it."

"But that will not be until next fall. What are you going to do for the next eight months?"

"I will study for the entrance exam."

The appearance of Albert shocked Hermann and it took him several minutes to adjust to it, but when he finally did he realized he was happy to see him and he knew he had missed him. He walked over to him and put his hand on his shoulder. "I'm not sure you've done the right thing son, but we'll see you through it."

Einstein thanked him, then offered to help at the factory. Hermann smiled at him affectionately as he accepted his offer.

Just before they retired for the evening Maja knocked on the door of the room where he was getting ready for bed. She smiled broadly as he opened the door. Her pajamas were too long for her and partially covered her bare feet. Her eyes sparkled as she said, "I'm so glad you came. It wouldn't have been Christmas without you."

Einstein knew he had done the right thing.

Einstein thoroughly enjoyed himself over the holidays and into the next year. In the spring they moved to Pavia and he made arrangements to help at his father's factory two days a week; most of the rest of the time he spent studying. He did, however, take weekends off to travel. Within a few weeks he had made several friends, and with one of them, Otto Neustatter, he hiked to several of the neighboring towns and cities. He visited museums and art centers in Milan, Padua, Pisa and several other cities.

It was early summer with raindrops splashing against the window. Pauline was working over the stove. Maja sat at the table.

"Albert has changed so much since he's been here," said Maja. "I've never seen him so happy. He was never this happy in Munich."

Pauline stirred some soup on the stove. "He should be happy," she said. "Nothing to do. He doesn't have to attend gymnasium. Hiking to this place and that. I think it's going to have to come to an end."

"But he's studying Mamma ... and he's helping Papa. Let him be happy."

Pauline frowned. "For a while," she said after a pause.

"Your father is having problems and it is difficult for us to have another mouth to feed."

The following day Pauline wrote to the Polytechnic in Zurich to find out about the entrance exams. She was worried. Although Albert was studying, he seemed casual and unworried, and he hadn't written the university on his own. She sighed as she thought about it, then she chuckled. He was, after all, only 15, and wouldn't be 16 until March.

She got a reply from the Polytechnic a week later. Her heart sunk as she read the letter: Anyone taking the entrance exams had to be 18 years old, and the rules were broken only in exceptional cases.

She presented the news at the dinner table. "The exams are in early October, and you will only be 16 1/2. You have to be 18," she said.

Einstein looked at her in confusion. "What can I do?"

Pauline turned to Hermann, but it was obvious he didn't have the answer. He looked back at her with a blank face, saying nothing.

"The letter says that the rules are broken in exceptional cases," said Pauline. She pursed her lips as her eyes narrowed. "This has to be an exceptional case." She paused. "I'll look into it."

Einstein's father no longer had the large study that he had had in Munich. His desk was now on one corner of the bedroom. It was cluttered with piles of papers and other paraphernalia. He called Albert into the room. Maja followed him in.

"Maja, please leave," said Hermann. "I want to talk to Albert alone. And close the door behind you."

Maja left.

Hermann sat down at his desk as Albert sat on the edge

of the bed. "Things aren't going as well in the business as we expected," he said to Albert. "We made several loans from your mother's family – the Koch's – and they're due." He paused. "I'm not sure how we're going to pay them off."

Einstein gave him a sympathetic look. "I'm sure you'll manage Papa."

Hermann toyed with some papers on his desk. "You've got to start thinking about what you are going to do."

"I am Papa. I will take the exams at the Polytechnic in the fall."

"But according to the letter your mother got you are too young."

"I'm sure that can be overcome. Mamma is going to write them."

Hermann leaned back in his chair. "What are you planning on becoming? What field have you chosen?" he asked cautiously.

Einstein stiffened; he had expected this, and he also knew that his father wanted him to go into engineering. He hesitated as he answered. "I would like to study philosophy –."

A look of pain came over Hermann's face. "What kind of philosophy?" he interrupted.

Einstein took a deep breath. "Kant, the great philosopher has published many interesting ideas about the origin of the solar system, the structure of the universe, the nature of matter and the meaning of forces. I would like to explore some of them."

"Rubbish, rubbish ... you cannot make a living doing that."

Einstein had anticipated this and he was ready. "I also have my mathematics and physics," he said quickly. "I am overflowing with ideas about electromagnetism and the ether. I would like to follow up on them ... do theoretical research. And I would teach mathematics and physics."

Hermann screwed up his face. "You have to make a living," he said firmly. "I think you should become an electrical engineer like your uncle."

"Electrical engineer!"

"Yes."

"I am interested in electromagnetism, but my interest in electrical machinery is only passing. I would not make a good electrical engineer."

"If you set your mind to it you would," said Hermann. He looked down to the floor for a few moments, then back at Einstein with compassion in his eyes. "I'm sorry son, maybe I'm being too hard on you, but I've had a lot of pressure on me lately. For my sake please tell me you will become an electrical engineer."

Einstein closed his eyes briefly as he thought it over. He didn't want to become an electrical engineer, but he didn't want to disappoint his father. Finally after several seconds he said, "Okay." He sat silent in his chair for several seconds, then said, "Papa, if I do that for you, will you do something for me?"

Herman straightened up in his chair. "Yes, of course," he said with a puzzled look on his face.

"I want to renounce my German citizenship and I need your signature."

"You what?" exclaimed Hermann in disbelief.

"I want to renounce my German citizenship."

Hermann leaned forward in his chair. "Why?"

"If I remain German I will have to serve in the German military and I don't want to."

Hermann ran his hand nervously through his thinning

hair. "I don't think it's advisable. You have to be a citizen of some country."

"When I go to Zurich I will become a Swiss citizen."

Herman sat in silence thinking it over. He was reluctant, but finally agreed to sign.

Over the next few weeks Einstein's enthusiasm for studying waned. He continued studying mathematics, physics and philosophy, but he was reluctant to study anything else. In addition to the promise to his father, there was still the problem of taking the exams. He might not be able to take them for another two years and the thought depressed him.

The Einsteins had a friend in Zurich and Pauline decided to write to him, hoping he might be able to help. Gustav Meier had been a neighbor in Ulm, where Einstein was born. He had moved to Zurich a few years after they had left Ulm, and they had kept in touch with him. She asked him to inquire about the possibility of getting around the age requirement for the Polytechnic exams. Meier wrote back that he had got her an appointment with one of the councilors at the Polytechnic.

It was a bright, clear day in July when Pauline and Albert got off the train in Zurich. Their appointment with the councilor was the next day. Albert looked over the city and the lake. The older section stretched upwards from the Limmat River, with the suburbs reaching around the shores of the Lake of Zurich in the distance. Vineyards and orchards covered the slopes of the hills that led down to the lake. It was a beautiful, picturesque city, framed by the snow-covered Albis Mountains in the background.

They took a carriage to the home of Meier.

Pauline decided that she would talk to the councilor

alone on the first visit. In early afternoon she took a carriage to the Polytechnic, or ETH, as it was commonly known. The main building was on the side of a hill, surrounded by other buildings and neighboring houses. Four stories high, it was an impressive building of brick and stone with countless long, narrow windows and gently sloping roofs over the various sections. Several large chimneys protruded from the roofs. The tops of the windows on the lower floors were rectangular; those of the upper floors were arched. The grounds were impressive with trees of all types dotting the lawns. Her appointment was for 2:00 P.M.

After asking for directions she made it to the councilor's office shortly before 2:00 P.M. She knocked on the large impressive door.

"Come in," sounded from within.

She opened the door and walked in. A man with dark, slicked-back hair and a mustache was seated at a desk.

"I had an appointment with you. I'm Pauline Einstein."

"Yes ... have a seat Mrs. Einstein. My name is Kurt Jost." He paused for her to get seated. "How can I help you?"

Pauline told him about Albert leaving gymnasium in Munich for health reasons. She showed him Dr. Talmud's note. "Although he is only 16, I hope the rules can be bent so he can sit for the entrance exams. He has spent several months preparing for them and I'm sure he can pass them."

Jost frowned. "We do not normally break the rules, Mrs. Einstein ... only in very exceptional cases."

"But Albert is a very bright student. He has always been number one in his class ... he's a prodigy."

Jost smiled. "Everyone's child is a prodigy in their mother's eyes," he said. "We get prodigies and geniuses here all the time and within a year they're gone."

"But I have a note from his mathematics teacher in Munich stating that he has studied higher mathematics ... including calculus. He has more mathematics than is required of students at gymnasium." She handed him the letter.

Jost unfolded it and began reading. Pauline watched his face carefully. As he finished he smiled; she was sure he was impressed. He looked at Pauline for several seconds. "How is he in other subjects ... besides mathematics?"

"He was number one in all his classes in Munich." She knew it was a lie, but a small lie wouldn't hurt.

Jost rubbed his chin. "All right," he said slowly. "I'll talk to the director, Dr. Herzog. Come back tomorrow at the same time and bring your son with you."

The following day Pauline and Albert were ushered into his office. Pauline introduced Albert, and they shook hands.

Jost looked him over carefully, then turned to Pauline. "I've talked to Dr. Herzog. What you are asking has not been done before. We have let some 17-year olds take the exam, but never anyone a young as 16." He turned to Albert. "According to the note from your teacher you were well ahead of your class in mathematics. What about other subjects?"

"I was also ahead in physical science."

"What about botany, zoology, languages and history?"

Einstein shrugged. "I was less interested in them, but I was always near the top of my class."

Jost continued quizzing him for several minutes, then said, "Wait here. I'll talk to Dr. Herzog again."

They sat in silence as they waited.

Within a few minutes Jost was back with an older man in a dark suit by his side. He was clean-shaven and had slightly receding gray hair. "This is Dr. Herzog," Jost said, introducing Pauline and Albert.

Jost quickly pulled up a chair for Herzog. "This is highly irregular and a definite breach of the rules," Herzog said to Pauline. He then turned to Albert, "But Mr. Jost assures me you are a highly unusual student."

Pauline smiled, sure now that her persuasion had convinced Jost.

Herzog sat staring at Einstein for several seconds. "Okay, we'll make an exception, but you will not be allowed to attend classes unless you pass the exam."

Einstein sighed as a smile came over his face. Pauline reached out quickly and shook Herzog's hand, thanking him.

With a little over three months until the exams, Einstein began preparing for them more seriously. But he still spent most of his time studying mathematics and physics. Every time he started studying botany or zoology his mind would begin to wander, and he would pick up a book on mathematics or physics and become sidetracked in it.

Pauline was busy over the stove as Einstein walked into the kitchen. He sat down in a chair and laid the book he had in his hand on the table. "There are so many things to memorize in botany and zoology," he said gloomily. "Why do you have to memorize them all?"

"Memorization is part of learning," said Pauline as she rolled out some dough.

"I'm not good at memorization."

"That's not true."

"It's difficult for me. I do not have a good memory."

"Nonsense," Pauline said sternly.

"I prefer to study mathematics and physics."

Pauline stopped what she was doing. "They are not your weaknesses. You have to concentrate on your weaknesses."

"Yes, but I'm going to study mathematics and physics if I go into electrical engineering. I won't be studying botany, zoology, languages and history."

"Still, you have to pass the entrance exams before you get in, and botany, zoology, languages and history are part of them." She put her hands on her hips. "I want you to study then. You cannot fail the exam," she said firmly.

Einstein looked sheepishly down at the floor. "I will."

Over the next few weeks Einstein continued studying, but he also continued to travel to nearby cities on the weekends. Pauline caught up with him one day as he was preparing to leave for Genoa. "You would do better staying home and studying," she said.

"I will be visiting museums. It will broaden my education."

Pauline sighed in disgust. "What about your work at your father's plant?"

"I'm not working until Tuesday and I'll be back by then."

"You're too young to go off on your own so much."

"Otto is going with me."

The sun struck the back of Einstein's jacket as he walked to his father's plant. It was warm with a few billowy clouds on the horizon. Flowers were beginning to bloom and the grass was starting to turn green. The plant was a long two-story box-like building that sat on a canal. The offices, which were one-story, were located at one end. Einstein picked up a stone and threw it into the canal, stopping briefly to watch the waves it created. Waves had always fascinated him. He rubbed his hands on his pants, then turned and walked across the street to the office building.

Jakob looked up as Einstein entered. He had been discussing a problem with one of the engineers they employed. "Maybe Albert can help us," he said, looking at Albert, then at his assistant. The assistant frowned in doubt.

Einstein walked up to them. "What do you want me to help you with?"

Jacob picked up a sheet with some equations written across the top and placed it in front of Einstein. He made a quick sketch on the clear part of the sheet. "The current in the solenoid is increasing linearly with time, so the flux is also proportional to time. If there are resistors here and here." He pointed to two points on the diagram. "What would the voltage be here, as a function of time?"

Einstein looked the diagram over as Jakob continued. "The problem seems to be this equation."

Einstein picked up the pencil and wrote down several equations. Soon the page was covered. As Jakob watched he solved the problem and handed the page to him. The answer was circled at the bottom.

Jakob's eyes widened in amazement as he looked at Einstein in admiration. "We've been working on this for days, and you solve it in ten minutes."

At that moment Hermann walked in. "Look at this," Jakob said to him. "Your son solved our problem as if it were nothing." Hermann smiled as he looked down at the page, but the marks on it meant nothing to him.

Although Einstein spent most of his time preparing for the Polytechnic exams, he refused to give up his studies on electromagnetism and the ether. The problem had become an obsession with him. What was the relationship between the electric and magnetic fields, and how did it relate to the ether? The ether had been invented a few years after the electromagnetic field was shown to have a wave nature. It was the medium that propagated the waves. Einstein was sure that the relationship between the electromagnetic field and the ether could be determined by studying potential states of the ether in a magnetic field, and by measuring the resulting elastic deformations and deforming forces.

During the summer of 1895 he wrote an essay outlining a program for studying it. Towards the end of the essay he wrote: "I believe that the quantitative researches on the absolute magnitude of the density and the elastic force of the ether can only begin if qualitative results exist that are connected with established ideas."

Einstein sent the five-page essay to his Uncle Caesar in Stuttgart. In the accompanying letter he mentioned that he would be attending the Zurich Polytechnic in the fall, but he didn't mention that he had to pass the entrance exams to do so.

Caesar was impressed with Einstein's letter and wrote back to Pauline praising him. Pride showed on her face as she sat back after reading the letter. But she was worried. Albert was studying, but she was sure he was spending little time on the subjects he was weak in. She cursed his stubbornness.

It was early October and the Polytechnic exams were only a couple of days away. Einstein packed for Zurich. When he saw his mother also packing he said, "You don't have to go. I can go alone."

"I'm going," she said firmly.

They caught the train the next day and stayed again at Gustav Meier's house. Einstein left early the following morning for the Polytechnic. The exam room was large, with desks scattered randomly over the floor. Oversized windows lined one side of the room, and there was a blackboard across the front of it. Several students were sitting at the desks when Einstein walked in. He took a seat and waited as others filed into the room. Finally a professor with a pile of exams under his arm approached the front of the room. "The first exam will be in mathematics," he announced as he began passing out the exams.

Einstein smiled; he was sure he would do well in mathematics, and indeed he had little trouble with it. The second exam that afternoon was on botany and zoology. This was the one Einstein was worried about. He read through the exam quickly and soon realized that he had not prepared himself well enough. He struggled through it, answering what he could, but uncertain in many cases. He had a hollow feeling in his stomach when he handed the exam in.

Later as he entered the front door of Meier's house Pauline jumped up from the sofa. "How did you do?" she asked.

"I'm sure I did well in mathematics."

"What other exams were there?"

"Botany and zoology."

These were the ones Pauline was worried about. "How did you do?" she asked cautiously.

Einstein shrugged. "I don't know."

The following day the morning exam was on physics and Einstein felt much better. As he handed it in he was sure he had done well. But that afternoon he had difficulty again with the language and history exams. His palms were sweating as he handed the last exam in.

Again, as on the previous day Pauline quizzed him

about how he had done, and again he was evasive.

"When will you know the results?" Pauline asked.

"They said they would be posted next Tuesday."

Meier offered to let them stay over until they got the results, and they took him up on the offer. As they waited Einstein visited several of the museums in Zurich. Outwardly, he appeared to be less concerned about the results than Pauline, who could hardly contain her nervousness. But inwardly he was worried.

On the day before the results were to be posted they got a message from Jost: "Please meet me in my office at 9:00 A.M. Tuesday morning."

Einstein and his mother arrived at Jost's office a few minutes before nine. They were surprised to find that Dr. Herzog was also in the office. Jost told them to take a seat. Herzog sat down on a chair near Jost's desk, facing them. He had a sheet of paper in his hand. Both Einstein and Pauline knew it was the results of his exams.

Herzog stared at the paper for several seconds, then looked up at them. Addressing Einstein he said, "You did very well in your mathematics exam," After a pause he continued. "Six out of six. It was the highest grade." He paused again, "And you got 5 1/2 on your physics exam."

Einstein sighed, but these weren't the subjects he was worried about.

Herzog pursed his lips. "Unfortunately, you didn't do as well on the rest of the exam. You got 3 1/2 in botany and zoology, 2 1/2 in languages and 4 in history. Overall your grades are not high enough to admit you to the Polytechnic."

Pauline's eyes widened in fright. "You mean he failed?"

"I'm afraid he did," said Herzog.

She glanced over at Einstein. His face was white and it

was obvious that he was in shock.

Herzog looked down at the page as if studying it, then up at Einstein. "We are impressed with your skills in mathematics, and your knowledge of physics. Professor Weber, who marked the physics section of your exam, was so impressed he was extremely disappointed when I told him you had failed the overall exam. He said to tell you that if you are staying in Zurich, feel free to audit any of his lectures." He leaned forward in his chair facing Pauline. "We would like to see Albert attend the Polytechnic."

Pauline's eyes lit up. "But if he failed the exam?"

"He'll have to make up his deficiencies," said Herzog.

"What should he do?" asked Pauline.

Herzog looked at Einstein. "You will not be able to attend this year, but there are many fine gymnasiums in the area. I would suggest you go to one of them and get a diploma. You will then be admitted next year without having to take the entrance exams again." He turned to Pauline. "It will only be a year and he is young."

"Do you recommend any particular school?" Pauline asked.

Herzog thought for a moment. "The cantonal school at Aarau is one of the best and it is not far away."

Chapter 4

1895 - 1896

Pauline made arrangements for Einstein to board with the Wintelers in Aarau. Jost Winteler was a history and philology teacher who lived across from the school with his wife and their seven children. He was to meet Einstein at the station.

The trip from Pavia to Aarau was one of mixed emotion for Einstein. Aarau was in the German speaking section of Switzerland, only a few miles from the German boarder, and Einstein was worried that the schools would be like those in Munich. But as the train approached Aarau his tensions eased. It was a beautiful town in a beautiful setting. In the distance were the magnificent Jura Mountains, and as the train began to stop he saw the maze of narrow streets rising from the Aare River below. The buildings were decorated with elaborate gables, wrought-iron emblems and delicately carved eaves, and every so often sunlight flashed from one of the many colorful windows. Surrounding the town was open countryside covered with trails.

As the train pulled into the station Einstein stared at the people standing on the platform. Jost Winteler was amongst them, but Einstein had no idea what he looked like. Pauline had written to him describing Einstein, and he hoped the description would allow Winteler to identify him.

As he stepped off the train he looked across the crowd. Somebody in a heavy dark coat grabbed the woman in front of him and began hugging her. He stared at the faces. Suddenly somebody tapped him on the shoulder.

"Albert?" said a tall man in a tweed suit.

"Are you Mr. Winteler?" Einstein asked.

"Yes."

Einstein reached out and shook hands with him. His first impression was favorable. Winteler's eyes seemed to sparkle and his voice was soft and friendly. He helped Einstein take his baggage to a waiting carriage.

"Did you have a good trip?" asked Winteler as he settled back in his seat.

"Yes," said Einstein, looking him over carefully.

"We live only a short distance away. It's very close to the school you will be going to. I'm sure you'll enjoy the school. The building that houses it now is old, but we will be moving into a new building in the spring, and it will be one of the best in Switzerland."

The prospect of a new building interested Einstein, but his main concern was the teaching methods and the discipline, and he was too embarrassed to ask about them so he listened and said little.

Winteler continued. "You will have your choice of either the gymnasium, which I'm sure you know, emphasizes the classics, or the technical and trade school. Both will allow you to attend ETH."

"I think I had enough of the gymnasium in Munich," said Einstein. "I would prefer the trade school."

"Your mother said you are going into engineering, so the trade school is perhaps the best choice."

Winteler leaned over and looked out the window of the carriage. "We're here," he said as the carriage began to slow down.

Einstein looked at Winteler's house as he stepped from the carriage. It was a large two-story white house with tall arched windows in the front and an elaborate entryway with double doors. The lawn was covered with leaves from the trees in the yard.

Einstein was introduced to each of the family members in turn. He was immediately attracted to Winteler's wife Pauline. She looked so delighted to see him he thought she was going to hug him. Of the three girls he was struck by the eyes and smile of Marie. She was the prettiest of the three, and Einstein soon learned that she was two years older than him. He was strongly conscious of her presence over the first few days, but said little to her, and was too embarrassed to do anything other than glance at her.

The morning after he arrived Einstein walked out the front door and looked across at the school. It was an old building; Winteler had told him that it was a hospital at one time. It didn't look appealing, but everyone was talking about the new school that was being built next door. Construction had begun on it, and Einstein could see that it would be a much larger building.

The following day Einstein signed up for his classes. To his dismay he found that singing, group exercising, and compulsory military training were on the curriculum, but he found he could substitute violin lessons for singing, military training was only required of Swiss citizens, and group exercising was optional. He registered for algebra, physics, French, German, Italian, chemistry, geology, geography and drawing. When he finished he took a tour of the building. The interior was as bad as the exterior; the hallways were narrow and dark, and the desks were old.

Most had been carved extensively with initials, and almost all were defective in some way.

Einstein's first class the following morning was physics. It was taught by the school rector, August Tuchschmid. Einstein sat at the back of the room as Tuchschmid filled the blackboard. To one side of it was a large cross-sectional diagram of an electric generator; it reminded Einstein of the diagram Jakob had shown him many years earlier. Tuchschmid was obviously tremendously interested in the subject and explained things with clarity and precision. Furthermore, he appeared to have an intense interest in electrotechnology, which pleased Einstein.

It was early in the evening, several days later. Einstein sat at the desk in his room. It was larger and considerably more comfortable than the room he had boarded at in Munich. Situated on the second floor, it had a window that looked out across to the school. As he sat hunched over a book Einstein heard music coming from the lower floor. Somebody was playing the piano. He sat listening to it for several minutes; it reminded him of his mother's playing, but was more delicate and tender. Finally his curiosity overcame him and he got up and went down stairs.

Marie was seated at the piano. She stopped playing as he came up behind her.

"Keep playing," he said. "It's beautiful. What is it?"

Marie blushed as she looked at him. "Chopin," she said as she resumed playing. "His Ballade in G minor."

Einstein listened attentively, then as she came to the end of the piece he asked, "Do you play Mozart?"

Marie nodded. "I'm working on his Sonata in C and I play his Sonata in A. I particularly like the last section … alla turka. I also play part of his Sonata in G."

"Who is your favorite composer?" Einstein asked.

Marie stared at him as she thought. This was the first time Einstein had a good, close-up look at her. She was even prettier than he thought. Her face had a look of childlike sweetness, and although shy, she was easy to talk to. Above all he liked her eyes; they were soft and serene, and when she smiled they sparkled.

"I love Beethoven's Sonatas," she said. "Particularly the Pathetique and the Moonlight, and I would like to be able to play the Appassionata, but it's very difficult. I also love Schubert."

"Ah ... a woman after my heart. I love Schubert too."

Marie tilted her head to one side and frowned slightly. Einstein saw that she had a dimple. "I find Liszt very hard, but I like Brahms."

"I play the violin. Would you like to play a duet with me?"

Marie smiled. "I'd love to."

"I have some music upstairs," said Einstein as he turned and rushed up the stairs to get it. Returning minutes later he handed it to her.

She looked it over. "I'd have to practice it a little first."

Within days they were playing the piece, usually with several other members of the family looking on. Einstein was beginning to feel much more comfortable around her now, and although her eyes still sparkled when she looked at him, she no longer acted shy. But with eight others in the house it was almost impossible for Einstein to be alone with her.

His first chance came the following weekend when Jost Winteler asked them to come along on a hike. Winteler was an enthusiastic ornithologist and hiked regularly in the nearby hills in search of rare birds.

"Bundle up. It will be chilly," said Winteler as Albert and Marie put their jackets on. It was a beautiful fall day with only a few clouds in the sky, but the air was cold. The group consisted of several students from Winteler's classes with Albert and Marie.

Albert and Marie started out at the tail end of the group and gradually fell farther and farther behind.

"This area is covered with flowers in the spring," said Marie. "You'll have to come up here then so you can see it."

Einstein smiled. "I will."

"Do you like hiking?"

"I didn't get much of a chance to hike in Munich, but I made up for it in Pavia. I hiked a lot, but I had to study for the exams at ETH."

"Don't you get bored studying so much?"

"It depends on what I'm studying. Mathematics and physics never bore me. Many things do, but you have to study them."

Marie walked over closer to Einstein, and he felt her take his hand. It was small and soft. For several minutes they walked without saying anything, then Einstein let go of her hand and put his arm around her. It was a new experience for him, but he liked it. She looked up at him and smiled.

Suddenly somebody from ahead yelled, "Hurry up you two." They rushed to catch up to the group, which had stopped. Winteler had spotted a bird and was studying it through his binoculars. Einstein watched as he began to whistle an imitation of the bird's warble.

A few days later the mountains were covered with their first layer of snow, and there were no more hikes.

Soon after he started classes Einstein joined the school orchestra. His love for the violin was as great as ever, and he played it at every opportunity. He sat back in his chair as the conductor told them to take a short break. Sitting next to Einstein was Hans Byland. Einstein had seen him in several of his classes but had never spoken to him. He felt Byland's eyes on him and turned toward him. "What are you staring at?" he asked coldly.

"You."

"Have you never seen a violinist before."

"You play beautifully," said Byland, ignoring the sarcasm. "There's a lot of feeling in your playing."

Einstein looked at him suspiciously, but he welcomed the compliment. "Thank you."

Byland continued to stare at him, which made Einstein uncomfortable. Finally he asked, "Do you count the music as you play?"

Einstein rolled his eyes. "Count? Heavens no. I don't need to. It's in my blood."

"I didn't mean anything by that," Byland said quickly. "I wasn't trying to offend you. I was just asking."

"No offense taken," said Einstein curtly.

That evening after dinner Winteler gathered everyone around the table. "This is our roundtable discussion," he said to Einstein when everybody was seated.

Einstein looked at him in surprise. "What do you discuss?"

"Anything ... what do you want to discuss?"

Einstein thought about his discussions with Talmud, but science didn't seem like an appropriate subject here. Winteler's area of expertise was history, literature and philology and he was sure the others knew little about science. Einstein shrugged, unsure what to say.

"With Albert being from Germany, maybe we could discuss Germany," said Anna, Winteler's oldest daughter.

"A good idea," said Winteler, looking towards Einstein. "What can you tell us about Germany?"

Einstein frowned. "I may be from Germany, but I don't have a good feeling for it."

"Why?" asked Pauline, surprised.

"Many reasons," said Einstein. "But the main one is that it is too militaristic."

"You don't approve of that?" asked Winteler.

Einstein shook his head. "Why does it need such a large army? Why does it need conscription? Why is there so much emphasis on the military? Who are they planning on attacking?" He was embarrassed by the aggressiveness in his voice, but if they wanted to talk about Germany, they were getting his opinion..

"Like any other country it needs to protect itself," said Anna.

Einstein made a face. "Who's likely to attack it?"

"France, or maybe Russia," said Marie.

"Germany would run over France in a few weeks, and as for Russia —"

"What about England and the United States?" asked Winteler.

Einstein was quiet for a few seconds. "England is a worthy foe, but United States is too far away."

"I'm surprised you talk that way," said Pauline. "Most Germans are very nationalistic."

"I'm not German."

"What do you mean?" asked Winteler.

"I renounced my German citizenship when I was in

Pavia."

Everyone looked at him in surprise. Finally Pauline asked, "Do you have no citizenship?"

"I'm planning on becoming a citizen of Switzerland."

She smiled. "I'm delighted."

"But Switzerland also has a army," said Marie. "You'll have to serve in it."

"I don't mind." He paused. "And there's another thing I don't like about Germany. If this conversation was taking place in Germany and anybody knew about it, we'd all be thrown in jail."

Several of the children laughed.

"I agree with you there," said Winteler. "And I sympathize with your point of view. Even now Germany is developing and experimenting with deadly nerve gases. They would be a horror in war."

Pauline got the children off to bed shortly after nine, then returned to the table. The discussion continued until after eleven. Einstein was pleased as he walked up to his room. Everyone was so liberal minded, and no one was offended when he spoke his mind.

Winter exams came in mid-December and as in the past, Einstein was not fully prepared for them. He got high grades in mathematics and physics and German, but most of his other grades were poor. Winteler frowned when he saw Einstein's report card, but he said nothing to him.

Einstein was embarrassed about his grades but he knew he had no one to blame but himself. What made it worse was that everything else in his life seemed to be going well. He loved staying with the Wintelers, and Marie had brought a new joy ... a new serenity and happiness into his life. He thought about her soft white skin, her slim waist and trim figure, her dimple and blue eyes. It was her eyes he liked most of all.

With exams over he was more relaxed. He sat at the dinner table watching everyone as he ate.

"I guess you're looking forward to going home for the Christmas holidays," Jost Winteler said to him.

Einstein looked over at Marie. He thought about Maja and his mother, and his home in Pavia. It had seemed delightful when he was there, but now he had something to compare it to. To him there seemed to be a lot more joy in the Winteler house, and he knew he would miss it and everyone in it if he went home. Turning to Winteler he asked, "Would you mind if I stayed here for Christmas?"

"We'd be delighted to have you," said Pauline before Jost could answer. Several of the children yelled in delight. Marie smiled. Einstein felt good; it was obvious that he was welcome.

The house was decorated for Christmas and a large Christmas tree was placed in the living room. It had snowed heavily throughout mid-December and the snow was now deep. Then just before Christmas it cleared up and turned bitter cold. The streets now had a thick layer of packed snow on them, and horse- drawn sleighs passed the house every few minutes. Ice and snow covered parts of the windows and at night the colors of Christmas were reflected in them.

The closer it got to Christmas the more excited the younger Winteler children got, and Einstein was soon caught up in the excitement, helping them make presents for one another. The family planned a party for Christmas eve. Several neighbors were invited, along with some of Winteler's students. Everything was decorated, including the front of the house. The younger children kept looking at the presents under the tree as people began to arrive.

Pauline had told them they could look, but nothing was to be opened until the following morning.

Einstein sat with Marie as everyone gathered in the living room. The children began playing games on the floor near the tree as the adults talked, ate and drank. Einstein did not know many of the people in the room, but with Marie at his side he did not care. She kept telling him who each person was as they entered the front door.

It was about ten o'clock when Pauline started getting the younger children off to bed. Marie turned to Einstein. "Do you want to go for a walk?"

"It's cold out."

"I know, but we can dress for it."

They put on heavy coats and slipped out the back door so no one would see them leave. It was dark, cold and clear outside. The stars shone brightly; they were like diamonds scattered over a sheet of velvet. The constellation Orion was prominent, high overhead, with Sirius to one side.

Marie looked upward as they walked. "Which one is the Christmas star?"

Einstein shrugged. "I don't know."

"You know so much about science I thought you would."

"I'm not sure anyone knows," said Einstein. "Wasn't it a miracle … if you believe in miracles?"

Marie leaned her head on his shoulder, and Einstein put his arm around her. "How far away are the stars?"

"So far away astronomers don't measure their distance in miles. They use light years … a light year is the distance light travels in a year. That star, " he said as he pointed at Sirius, "is over eight light years away. And some of the others are hundreds of light years away."

They stopped and sat down on a bench. A sleigh with

bells jingling passed them. They stopped talking as they watched it.

Einstein looked again at Sirius. "A light beam from that star takes eight years to get to us, yet it travels at 186,000 miles every second." He moved his head slowly from one side to the other as he marveled. "It's hard to believe."

"How far away is it in miles?"

Einstein thought for a few seconds. "About 50 trillion."

"Gee, that's a long ways."

"Did you know that light is a wave?" Einstein asked.

"I never thought about it. What kind of a wave?"

"Just like the waves that come when you throw a rock in a pond."

She looked confused. "How is that possible? There's no water in space."

"That's true, but scientists believe there is a medium for propagating the wave. They call it the ether."

"Do you believe it?"

"I'm not sure. There are a lot of problems with the idea."

She looked at him in admiration. "You know so much. Where do you learn it all?"

"From books." Einstein looked up again at Sirius. "The wave from that star stretches all the way to us. I've been wondering lately what it would be like to jump on the wave and travel along with it."

"How could you do that?"

"You couldn't ... but it's an interesting thought."

She moved closer to him on the bench. He looked at her soft white face, wondering what it would be like to kiss her. He had never kissed anyone except his mother, and that wasn't really a kiss. He had not even thought of kissing a girl before Marie had come along, and he wasn't sure he had the nerve to try kissing her.

He pulled her closer to him. She looked at him in the darkness, then leaned her head on his shoulder. Her hair smelled clean and fresh, and there was a slight aroma of perfume about her. He moved his cheek slowly toward hers as his heart pounded. Her skin felt warm in the cold night air. He kissed her lightly on the cheek, then moved toward her lips. She turned, meeting him, as her arm came around him.

Einstein ran down the stairs as the mail came. He had seen the mailman approaching from his second story window, but Pauline beat him to it. He waited as she sorted through the letters. "There's one for you. It must be from your mother," she said handing him the letter.

Einstein ripped it open and began reading. "My father is coming to visit us," he said excitedly. "He has some business in Zurich and wants to come and see me."

"We'd be delighted to have him," said Pauline.

Two days later Hermann arrived by train. Einstein and Jost Winteler met him. Einstein was excited and talked nonstop throughout the carriage ride to the Wintelers. As they got out he pointed to the old school and told his father about it, then he told him about the new school which was now partially finished. Once inside the house, he rushed him up the stairs to show him his room.

Later that evening Jost and Hermann were alone in the living room. "Thank you very much for the hospitality," said Hermann, "and thank you for making such a delightful home for Albert. I can see that he is enjoying it immensely."

"We are enjoying him. He's a bright boy." Winteler fell

silent for a few seconds as he looked down at the floor. "I'm worried, though," he said, raising his eyes. "His grades … the grades he got in December weren't very good."

"I hadn't heard."

"He's intelligent. There's no doubt about that. You only have to talk to him for a while to see that. We discuss many things over the dinner table, and his insights always amaze me. He got high grades in mathematics and physics, but some very low grades in other subjects. I don't understand. He did have several things against him: he came in at midyear, and therefore had to catch up, and he's a year younger than everyone in his class."

Hermann chuckled. "I don't think it's anything to worry about. It's typical of Albert. He did the same thing on the Polytechnic exams, and at Munich he always got a few low grades. But in most classes he was near the top. He's capable of doing it … if he puts his mind to it. His problem is that he hates studying anything that doesn't interest him. In the end I'm sure he'll come through."

"I feel the same way, but he has his heart so set on going to ETH I would hate to see him fail."

"I'm sure he won't."

Winteler smiled as he looked down. "Another distraction is Marie. They appeared to be fascinated with one another, and they enjoy one another so much I don't want to discourage it."

"She's a very pretty girl. I can understand his attraction." He laughed.

When the spring break came in late March Einstein was finally eager to go to Pavia. It had been a long time since he had seen his mother and Maja and he was beginning to miss them.

The train stopped in Pavia late in the afternoon. Pauline and Maja met him at the station. Einstein knew they would ask him about Marie, and the questioning began soon after they were in the carriage. Maja was even more curious than Pauline.

"Tell us about Marie," Maja asked.

"What's there to tell?"

"What does she look like?"

Einstein smiled. "Why do you want to know?"

"Come on. Tell us. Papa said she's pretty."

"She is ... she's very pretty."

"Is she intelligent?" Pauline asked.

Einstein leaned his head to one side. "I wouldn't say she's an intellectual, but she's bright."

"How old is she?" Maja asked.

"Eighteen."

"She's older than you," Pauline said in surprise.

"Yes, but in many ways she seems younger."

"Are you in love with her?" Maja giggled.

Einstein smiled. "What do you know about love?" He looked down at the carriage floor briefly. "I might be. I like her a lot."

"At 16 it's just puppy love," Pauline said with mock disgust on her face. "You're too young to be in love."

"Maybe you're right. I've never been in love before. I think I'm in love but I don't know for sure."

Maja started giggling again. "Have you ever kissed her?"

"I might have," he said, looking at her as if to say, "mind your own business."

"I bet you don't even know how to kiss," Maja teased.

"I do too."

"Show me."

"That's silly. How can I do that?"

While Einstein was at home Pauline had him play his violin for several of the ladies of Pavia. She was particularly proud of his playing and loved to show him off, but Einstein soon tired of the charade, and within a few days he was pining for Marie. The feeling of melancholy became worse when he got a letter from her.

That evening he sat at the kitchen table with a pen in his hand. "Strange how I thought Pavia was such a great place," he said to himself. "It's nothing compared to Aarau." He dipped the pen in the ink and began to write:

Dear Marie:

Many thanks for your cheery letter, which made me endlessly happy... I was made to realize to the fullest extent, my angel, the meaning of homesickness and pining. But love brings much happiness ... Only now do I realize how indispensable my dear little sunshine has become to my happiness... You mean more to my soul than the whole world did before.

Kisses and all my love
Albert

As the end of the holidays approached Einstein was anxious to get back to Aarau. He was homesick for the entire family, but in his heart he was most anxious to see Marie. He spotted her out of the window of the train as it pulled into the station. The sight of her made his heart jump. He rushed down the aisle as the train came to a stop.

Jost, Pauline, Marie and Anna were all there to meet him. He hugged each of them in turn. Marie had a pink dress on that showed off her trim waistline; it had a high collar and she had tied a small bow in the front. He had never seen her more radiant and pretty.

While he was gone the new school was completed and it was now ready for dedication and occupation. It was a magnificent building: four stories of brick and stone with a large arched entryway and a clock tower on the top. But what was of most interest to Einstein was what was inside. The laboratories were amongst the best in Switzerland, each equipped with the latest equipment. In the physics lab there were alternating current motors, dynamometers, batteries, several modern galvanometers, and other meters. The adjoining shop was equipped with a milling machine, a lathe and an endless variety of tools.

The dedication took place on April 26, 1896. It was a beautiful spring day and the ceremonies were held out of doors. Several speakers were brought in. Einstein's physics teacher, the rector of the school, gave a long speech on the value of education and the higher principles of life.

They had barely moved into the new school when the Mayday celebrations were upon them. Einstein was in the middle of an experiment on the last day of April when Byland approached him. "Einstein, we need you," he said.

Einstein looked up. "What for?"

"You've heard about the big fight on Mayday, haven't you?"

Einstein shook his head. "No."

"The students at the school challenge the men of the town to a sword fight. We get all dressed up for it."

"You mean they actually have a sword fight?"

"Yes."

"What kind of swords do they use?"

"Fencing swords."

Einstein frowned. "Somebody could get hurt ... besides,

it sounds silly to me."

"It is silly, but it's a heck of a lot of fun, and nobody gets hurt ... at least not badly." He paused. "We need one more person. Would you be willing? I have an outfit and sword for you."

Einstein was hesitant, but after thinking it over for a few seconds he said, "I guess it won't do any harm."

"Good."

"But I still think it's silly."

Byland shook his head. "Einstein ... I don't think you'll ever change."

Einstein looked at him in surprise. "Change? Why should I?"

"Forget it."

The fight was to take place the following day. Dark clouds covered the sky as the two sides got ready for the battle. As the first few drops of rain began the two sides ran for cover in a building at the edge of the field.

Einstein ran down the street, rushing to get to the field on time. He was dressed in a black fencing outfit, and on his hat was a skull and crossbones. When he got at the field he was surprise to find that no one else was there. As he looked around the few drops of rain turned to a downpour. He headed for the nearest building, but by the time he got there he was soaked.

Byland and several others were in the building. Einstein was shaking the rain off his hat when Byland walked up. "The fight was called off," he said.

"Called off ... nobody told me."

"You were late." Byland shook his head as he looked at him. "You're quite a sight, Einstein," he said with a chuckle.

"Go ahead and laugh," said Einstein sarcastically. "And you can get somebody else next time."

Einstein did much better on the spring exams in late May. With the exception of French and technical drawing, his grades improved significantly. He did well enough in geology, in fact, to be selected as one of 20 students asked to accompany Professor Muhlberg on a field trip to Mt. Santis, near Appenzel in the eastern part of the country.

Einstein was anxious to tell Marie. She was ironing when he walked into the kitchen.

"I'm pleased for you," she said when he told her. "And I have some good news. I got a job teaching at Olsberg in the fall. It's only temporary but it will give me some experience."

Einstein stared at her for several seconds. "I'll be in Zurich in the fall."

"We can write to one another," said Marie.

"But we might not see one another very often."

"We can make arrangements."

"It will be difficult."

She shook her head. "Love overcomes everything, sweetheart."

He quickly changed the subject. "I'll be gone for several days on the field trip. We leave the day after tomorrow."

"Will you write me?"

He nodded.

Two days later Einstein assembled at the station with 19 other students and Professor Muhlberg. Despite his curt, abrupt manner, Einstein liked Muhlberg; furthermore he admired him as a scholar. Muhlburg had published extensively over the years and was an active researcher.

The train stopped at Toggenburg; Einstein got off,

bought a postcard and wrote to Marie as he had promised. He apologized for what he had said a few days earlier, and told her that he wanted to correspond with her when he was in Zurich, and he hoped to see her frequently after they were separated. It was still a couple of months before they would be split up and they should make the best of it until then.

The train continued on to Appenzel. Einstein marveled at the scenery as they approached the station. Rolling hills covered in a carpet of green, varying from lime to dark moss, broken by clumps of trees and fairy-tale farmhouses lay on both sides of the train. Cows were grazing in many of the fields, and looming in the background was the Aplstein range and 8,141 foot Mt. Santis. The top of Mt. Santis was shrouded in fog.

It was damp and cold for June when Einstein and the others got off the train. They had hoped for good weather but dark, ominous clouds hovered overhead as they made their way to the base of Mt. Santis. Much of the time they could barely see the majestic mountains that surrounded them. Muhlberg assembled them and gave them a few words of advice when they reached the trail at the base of the mountain.

The trail was steep and Einstein had done little hiking during the spring and was not in good shape. He could feel his heart beating with each step. Every so often Muhlberg would stop and point to a geological formation, then ask one of the students a question about it. Einstein welcomed the rests.

Einstein and Byland were frequently near the tail end of the group and they managed to avoid being asked any questions. But late in the afternoon Einstein heard his named called. Muhlberg was pointing at a formation below them. "How does the strata run there? From below upward of vice versa?" he asked Einstein in his usual curt manner.

Einstein stared at it casually for a few seconds. "It's pretty much the same to me which ever way it runs, Professor."

Several of the students laughed. Byland rolled his eyes.

"Albert, that will be enough of that," said Muhlberg sternly. He glared at Einstein for several seconds then turned to another student.

On the upper part of the mountain they encountered considerable fog and finally the trail was covered with snow. Several of the students slipped on the wet snow and Muhlburg cautioned them to stay away from the edge.

"Somebody has fallen," echoed up the trail as thick fog swirled around the line of students.

Muhlberg came rushing back. "What happened?"

"Somebody's over the edge," said one of the students.

"It's Einstein," said another.

Muhlberg made his way cautiously to the edge and looked down in the fog. He could see Einstein was on a snow-covered ledge several feet below. "Einstein, stay where you are. Don't move."

Einstein looked down in the fog. All he could see was a blanket of white, but he knew if he slipped it would be the end of him. His heart pounded.

Byland got down on his stomach on the snow and reached down toward Einstein, but his arm wasn't long enough.

"Does anyone have a rope?" Muhlburg yelled.

No reply.

"We don't need a very long one."

"Would my sock do?" asked one of the students.

"Maybe," said Muhlberg.

The student took it off and handed it to him. Muhlburg pulled on it, testing its strength. He started to lean over.

"Here, let me do it," said one of the taller, stronger students.

Muhlburg handed the sock to him. The student got down on his stomach and eased himself over to the edge. He lowered one end of the sock to Einstein. "Have you got it?"

"Yes."

"Can you pull yourself up?"

"I don't know. I'll try."

"Just get your hand up to mine. I'll pull you up" Then in a stern voice. "But don't let go. You'll go right over the edge."

Einstein didn't have to be told. "I know," he said in a weak voice.

He pulled himself upward on the heavy sock, trying to keep his mind off what would happen if he let go. His arms ached, but he placed one hand over the other and gradually got to the top of the sock.

"Hold on tight, but give me one hand."

Einstein held his breath as he let go with his left hand and thrust it upward. The student grabbed it and began to pull. With the additional force on him, however, he began to slide closer to the edge. "Hold onto me," he yelled. Several students grabbed him.

Slowly he pulled Einstein up the icy edge. Einstein's face was white when he was finally pulled to safety. He was panting so hard he could barely say, "thank you."

"That was a narrow escape," said Muhlburg. "I think this is as far as we should go."

Einstein was grateful for the help he received. He knew

he could have been killed, and for the next few weeks there was little sign of his usual cockiness. By late summer, however, his mind was in a whirl of confusion and worry. Final exams were approaching and he was worried about letting his parents and the Wintelers down. He had to do well, which meant he had to spend less time with Marie.

He cursed to himself as he thought about the promise he had made to his father. He didn't want to become an engineer; teaching was a much better profession for him. He admired Winteler, Tuchschmidt and Muhlberg and was convinced that they had much better lives than either his father or uncle. They had considerable freedom to do what they wanted, and they weren't always worried about business and money.

He was in a bad mood as he sat at his desk, hunched over a book. A light knock came on his door.

"Come in."

Marie entered; she looked downcast as she walked over to this desk and sat down in a chair beside him. "I'll be going to Olsberg soon."

Einstein stared at her, saying nothing. Finally, "We're each going our own way."

"It doesn't have to be that way."

Einstein looked at her almost in pity. She had brought him so much joy, but he didn't know what to do. He couldn't spend all his time thinking about her. There was too much to do.

"I'll write to you," she said after a short silence.

"Maybe it's best that we don't write to one another."

A shocked look came over her face. "Why?"

"It's just for the best."

Tears started to form in her eyes. As she went to speak her voice cracked and she said nothing for several seconds.

Finally: "Albert, I love you."

Einstein put his arm around her. "I love you too but what else can we do?"

She took a handkerchief from her sleeve and wiped her eyes.

"I've got to concentrate on my exams. I can't afford to fail them." said Einstein.

"I won't bother you, I promise." She dabbed her eyes again. "I don't want us to part. Will you write me in Olsberg?"

"Yes, I promise."

Einstein took his final exams on September 18, 19 and 20th and defended them on September 30. He received the highest grade in his class in both mathematics and physics, and he did well in chemistry and geology. His German essay on Goethe's plays also earned him the highest grade in his class. But he was near the bottom in French and drawing.

Chapter 5

1896 - 1898

In the summer of 1896 the Einstein brothers business failed for the second time. They had set up the business about a year earlier with great expectations, but in their effort to succeed they had overbid several projects and found themselves heavily in debt. They discussed their problems with Garrone and soon saw that they had no alternative but to liquidate. But even with the liquidation Hermann was still far in debt, and coming so soon after the business failure in Munich it left him despondent.

Einstein was home for the holidays when Pauline made the announcement at the dinner table. Brushing away a hair that had fallen down across her face she said, "Albert, Maja ... I have some bad news."

Einstein knew there were problems with the business and he had been expecting something, but he had continued to hope that things would turn around.

"Your father has had some business problems and we will have to move," said Pauline.

Einstein groaned as he thought about the Polytechnic. What kind of a strain would it put on them? He looked over at his father, feeling sorry for him as he several times in the past. It had to be hard on him, and he had to be embarrassed. As in Munich when the business failed, Hermann said little; he sat with his head down, staring at his plate.

"Where will we be going?" asked Maja.

"We will be moving back to Milan."

Einstein looked at his mother in surprise. "Why Milan?" he asked. A move back to Milan seemed to make little sense.

"Your father is going to set up a new business in Milan."

Einstein bit his lower lip as he looked over at his father; despite the sympathy he felt for him he was annoyed. Two business ventures had failed, why try again? "What about Uncle Jakob?" he asked. "Is he going into business with you again?"

Hermann said nothing for several seconds, then he shook his head. "No, he has taken a job at an electrical company."

Well, at least he's smart, thought Einstein. "You should get a job too," he said. "Another business would put too much pressure on you."

His father shook his head. "No, I think I can succeed this time. Things will be on a smaller scale ... no large overhead. I'll only be selling generators and electric motors."

Einstein protested. "You were sure you would succeed last time."

But Hermann was adamant, even though he knew there was a serious problem, a problem he had not yet mentioned to Pauline. As he sat staring at his plate he knew he had to bring it up. He glanced at Pauline, then turned to Einstein. "Everything I have will have to go into the business." He hesitated. "We won't have any money to pay for your education at the Polytechnic."

Pauline had been expecting this and she was annoyed. She let her feeling be known. "I've gone through a lot to see that Albert goes to the Polytechnic and I'm not going to see it denied to him," she said in an angry voice.

"I'm sorry," said Hermann apologetically.

Pauline pursed her lips in grim determination. "I'll write to Julie Koch and see if she can help."

Einstein had never cared for his Aunt Julie. To him she was pompous and vain, but if she would help he would gladly accept it.

Several days later Pauline got a letter back from Julie. She promised to give Einstein an allowance of 100 francs a month. Both Pauline and Einstein were relieved when they heard the news.

It was the second week of October. The skies over Zurich were grey and the air was chilly. Einstein stepped down from the carriage and looked at the house on Unionstrasse. He was dressed in a grey suit with a vest and tie. It had rained the night before and he was careful to avoid the puddles on the street. He took out the advertisement he had clipped from the newspaper and checked it as the driver watched him. Satisfied it was the right address, he surveyed the front of the house. It was three-stories high, with long, narrow windows in front and a steep roof, similar to the houses around it. Einstein turned to the driver and told him to wait, then he walked up to the door and knocked. Within minutes a short, overweight woman with grey hair appeared.

"Frau Hagi?" Einstein asked.

She nodded cautiously. "Are you looking for a room?"

"Yes," said Einstein.

"I have two left. You can have your choice." She led him up a flight of stairs to a room in the back of the house. It was small, with a bed along one wall and a narrow desk along the opposite wall. The paint on the baseboards was chipped in places and the wallpaper was old, but there was a large window at the end of the room that made it bright. Einstein thought that if he took the room he would move the desk under the window. Through it he could see the Polytechnic a short distance away.

He asked to see the other room before he decided. They were similar, but he liked the view from the first room better, and decided to take it. He got his luggage from the carriage and paid the driver. Then, after paying the landlady for the first month, he carried his luggage up the stairs. As he unpacked he thought about Aarau and the Wintelers; he missed them, and he also missed his parents, but his dream of attending the Polytechnic had finally come true and he was looking forward to it.

He arrived at the Polytechnic early the next morning. About a thousand students were registering so the halls and grounds were crowded. Almost all the students were male and most were registering for engineering. Einstein was a year younger than most of them; he was still six months short of his eighteenth birthday, and eighteen was the legal age for entrance to the school. After asking directions several times Einstein found Department VI – the school for teachers. He registered in section VIA, the mathematics section; it included mathematics, physics and astronomy. Although his major was physics he would be able to teach both physics and math.

One of the professors helped him work out his study program. He would be taking only mathematics classes the first semester. His first class in physics wouldn't come until the second semester. When he finished registering he walked out on the terrace and looked down at the city center below. Through it ran the broad, slow-flowing Limmat River. Several low bridges spanned it, and as he watched a small boat passed under one of them. It was mid-morning and the traffic was relatively heavy; he could see several carriages on the nearest bridge. The spire of one of the highest buildings on this side of the river caught his attention; near the top was a clock and it made him think of how time was treated in physics. To most, time was a simple concept that deserved little thought, but he was convinced that it was much more complex than people realized.

As he came out of his reverie he thought about what he would do over the next few days. Classes would not start for two days so he would have time to familiarize himself with the city. Although he had visited several museums and parks when he was here a year earlier, be was still unfamiliar with most of the city. He decided to spend the next two days exploring it.

His first class, a mathematics class taught by Professor Hurwitz, came the following Thursday. Einstein had studied calculus earlier on his own and was familiar with most of the material Hurwitz lectured on. It gave him a feeling of confidence; he was sure he would have no trouble with the class. Looking around at the other students in his class, he was surprised to see that one was female. She had dark hair and a dark complexion; Einstein was sure she wasn't Swiss, but he was unsure of her nationality. Dressed elegantly in a new suit, she had a firm chin, penetrating dark eyes and appeared to relatively attractive. She made him think of Marie. He was uncertain now of his relationship with her, but she was still writing to him, and he was writing back.

When the lecture was over one of the students approached him. "Are you in section VIA," he asked.

Einstein nodded.

The student extending his hand. "Me too. We'll be seeing a lot of one another. I'm Marcel Grossman." Einstein looked him over as he shook hands. He had dark wavy hair, a small neatly trimmed mustache, and he was slender and wore an expensive suit. "Have you met the other students in our class yet?"

"No," said Einstein.

"That's Louis Kollros," Grossman said, nodding towards someone exiting the room. "He's majoring in math. The guy that was sitting next to him was Jakob Ehrat. He's also a math major as I am."

"Have you talked to them?"

Grossman shrugged. "Just introduced myself."

"I see there's a girl in the class."

"I think she's a physics major. Her name is Mileva Maric, but I don't know anything about her."

Einstein and Grossman talked for several minutes as they exited the room and walked down the hall. Einstein liked him immediately. He was friendly, but not overbearing and he appeared to have a strong love for mathematics. Strangely, though, he looked upon physics as a necessary evil, something he had to take – like medicine – but he wasn't sure it was ever going to do him any good. Einstein chuckled at this; in many ways he regarded mathematics in the same way. Grossman told him he had gone to school in Basel, and that his father manufactured agriculture equipment. Einstein smiled to himself as he left Grossman; he was sure it was the beginning of a strong friendship.

It was two weeks before Einstein spoke to Mileva for the first time. She had smiled at him when he bumped into her, but it was apparent to him that she was reserved, and he was curious about her. It still seemed strange seeing a girl in mathematics and physics. He followed her down the hall after class, walking directly behind her, so close he could smell the perfume in her hair. She had a slight limp, but it didn't bother him.

"Hello," he said as he pulled up beside her.

She glanced over at him, but said nothing.

"I'm Albert Einstein."

"I know."

"It must be difficult for you ... being the only girl in the class."

"I'm use to it."

"Where are you from?"

Mileva slowed her pace as she cocked her head. "Southern Hungary. I'm Serbian, but I graduated from the Young Ladies College here in Zurich."

"Why are you studying mathematics and physics – if I may ask? They seem like difficult subjects for a girl."

Mileva looked over at him and frowned. "Why are they more difficult for a girl than a boy?"

Einstein blushed as Mileva continued. "My highest grades were in math and physics at the Young Ladies College, but my real interest at the time was medicine. I attended the summer session of the medical school at Zurich University, but by the end of the semester I knew it wasn't for me. I had always done well in math and physics so I decided to teach them. That's why I'm here."

Einstein smiled. "Well, it's interesting having a girl in my classes. There were all boys in my class in Aarau."

"You're from Aarau?"

Einstein shook his head. "My parents are living in Milan, but I'm originally from Munich." He went on to tell her about dropping out of gymnasium, then reluctantly he told her about taking the Polytechnic entrance exams and failing. "What do you do when you're not studying?" he said finally, trying to change the subject.

Mileva thought about it. "I read a lot ... and I love music."

"Do you play an instrument?" he asked quickly.

"I play piano and sing."

He smiled broadly as he looked at her more closely. Her dark, shining hair was swept back in a large roll. It was strange how different her eyes were than Marie's; they were dark and penetrating, and he was sure they missed little.

He continued thinking about her as he walked back to his room. There was a letter from Marie waiting for him on the small table inside the door. He carried it up to his room without opening it. Unconsciously he began comparing Mileva to Marie. Marie was prettier, but Mileva had so many things that appealed to him. She was intelligent, she understood physics and math, she played the piano, and in addition, she sang.

He sat down on his bed, opened the letter and began to read. She scolded him for taking so long to reply to her last letter. He had complained in his last letter about the problem of getting his laundry done, and she offered to do it. He could send it to her. He lay back on the bed as he finished the letter. He was still confused about how he felt about her; he liked her, but he was sure he didn't love her.

It was late November and the tops of the distant mountains were snow-covered, but there was no snow in Zurich. It was cold, however, and had been foggy every morning for the past week.

Einstein had agreed to meet Grossman at the Cafe Metropol on the banks of the Limmat River. He sat at a table looking idly at the river as he waited for him. Within minutes Grossman came through the front door. He had an overcoat on over a dark suit. Einstein waved him over to the table.

A waiter came up as Grossman took off his coat and

seated himself. Einstein ordered coffee and a pastry. Grossman ordered coffee.

"What do you think of the Polytechnic now that you've been here for a while?" asked Grossman as he relaxed back in his seat.

Einstein looked out through the window at the river as he thought about the question. He still had a strong aversion to teachers as a result of his experience at Munich, but his year at Aarau had moderated his feelings. Tuchschmidt, in particular, had changed his attitude, but he had not yet had time to judge the professors at the Polytechnic. "The teachers are very good ... at least the ones I've seen so far. Hurwitz is excellent, but I'm fairly familiar with most of what he has covered so far."

"You studied calculus on your own before you came here?"

Einstein nodded.

"What made you decide on physics?"

"My real interest is theoretical physics." He paused as a look of intensity came over his face. He enjoyed experimental physics, but he knew he was better at theory than he was at experimental physics, and it was much more logical for him. "I have a feeling big changes are coming in physics ... perhaps a revolution, and I would like to be part of it."

"I feel the same way about mathematics. I'm sure there's still a lot to be learned in it too." He paused. "Did you ever think of going into mathematics?"

Einstein shook his head. "I suppose I thought about it, but from what I've seen there's a big difference between math and physics. Mathematics is split into so many specialized areas, each of which could take a lifetime to master. Physics is not yet at that stage. There are many more opportunities in it."

Grossman leaned back in his chair and smiled. "But mathematics is a more exact science than physics. Much of physics is based on hypothesis."

"That may be true," said Einstein as he took a bite from his pastry. "And indeed many of the hypotheses may eventually be proved to be wrong. Maybe that's why I feel there's more opportunity in physics."

"It looks like it might be a while before you get a taste of physics. Herzog's dynamics class comes next semester, but it will be strongly oriented towards engineering. You'll have to wait for Weber's class in the third semester."

Einstein took a sip of his coffee. It was annoying to him that he would have to wait a full year before getting to a physics class. Mathematics was important in the understanding of physics, but to him it was secondary, and taking a full year of mathematics before getting to physics made little sense. "I know," he said, "but I'm doing a lot on my own. Sometimes it's better to dig out the material on your own, rather than have it fed to you in a lecture. You get a better understanding of it.

"I don't have any problem with that," said Grossman, "as long as you don't neglect the lectures." He laughed. "There are the finals to pass."

Einstein approached the house with anticipation. His violin case was tucked under his arm, and it was now several weeks since he had played for anyone. One of the students had told him that there was a musical get-together at the house of Selina Caprotti each Saturday afternoon. Musicians and people interested in listening to music gathered to enjoy music. Einstein stopped in front of the house and looked it over. Despite the November chill one of the upper windows was open and he could hear a piano. The building was a grey stone, three stories high, with large windows in the front, each surrounded by dark brown shutters. Two chimneys stood on the steep roof.

Einstein knocked on the door and was admitted by an attractive woman with dark hair pulled back in a bun behind her head. She smiled when she saw the violin under his arm. She introduced herself as Selina Caprotti and led him into a large room with a grand piano at one end. Several people sat on couches and chairs around the room. The pianist stopped playing as Einstein entered; Selina Caprotti introduced him to everyone in the room, then asked him to play. Several people clapped when they heard the invitation.

Einstein was delighted. He walked up to the pianist and asked her if she was familiar with Mozart's Sonata in A. She looked over the music that he handed her and agreed to play.

The music encompassed the room as everyone watched and listened to Einstein and his accompanist play. When they finished there was loud clapping and several cheers. Einstein was pleased; as he was putting his violin back in his case when he was approached by a short man with curly black hair and a beard.

"I enjoyed your playing," said the stranger.

"Thank you," replied Einstein.

"I'm Michele Besso, and I'm a great fan of Mozart."

Einstein studied him. He could see that he was several years older than most of the Polytechnic students, nevertheless he asked hesitatingly, "Are you a student?"

"Yes, I'm in engineering ... mechanical engineering."

"From around here?"

Besso smiled broadly. "I grew up in Italy, but I was born

not far from here, so I'm as Swiss as most people around here, although I may sound Italian." He laughed. "What about you?"

"I'm a physics major from Germany."

"I've always liked physics," said Besso. "But mechanical engineering seemed a little safer for me. I don't think I have enough imagination for physics. What area of physics are you planning on specializing in?"

"I'm interested in theory ... electromagnetic waves, atoms and so on."

Someone began playing Beethoven's "Moonlight Sonata" on the piano and both men looked around. The pianist was a young woman; they stopped talking as she played.

"Beautiful piece," Einstein said as she finished.

"You said you were interested in theory."

Einstein nodded. "Have you ever heard of the ether?"

Besso rolled his eyes. "It was mentioned in a class I took, but I'm afraid I don't understand it.

"I don't think anybody really understands it," said Einstein. "I'm seriously bothered by it. It seems strange that a substance that permeates the entire universe can only be detected indirectly through its ability to propagate waves."

"It's invisible, isn't it?"

Einstein nodded.

Besso stared at him in admiration. Finally he extended his hand. "It was nice to have met you. We'll have to talk again."

Einstein watched him as he left the room. An odd fellow, he thought, but friendly.

Einstein continued to get letters from Marie but he finally stopped answering them. In the meantime he was in classes every day with Mileva, and his interest in her continued to increase.

"What's your hometown like?" Einstein asked as he walked down the sidewalk with her.

"I didn't live in a town. I lived on a farm in the country."

"Well, what is it like?"

"Flat. Lots of corn and sunflowers. And there's the beautiful Danube River. Novi Sad's the nearest town. Our farm is at Kac."

"Any special friends back there?"

Mileva chuckled. "What kind of friends? A boyfriend?"

"Yes."

"Doesn't every girl have boyfriends." She smiled smugly.

"I don't know. Do you?"

"I might. What about you?"

Einstein shrugged. "I had a girl in Aarau, but it's over now."

"What was her name?"

"Marie."

"Was she pretty?"

"Very."

Mileva looked at him with a sparkle of jealousy in her eye.

With the Christmas break Einstein went back to Milan for Christmas. His mother met him at the station; Maja was in school and unable to come. During the carriage ride to their house Pauline asked about school, and seemed pleased when he told her he was working hard and having little trouble. Then unexpectedly she asked, "Why aren't you answering Marie's letters?"

Einstein looked at her in surprise. "How do you know

I'm not?"

"She wrote me."

Einstein sat up in his seat. "Why would she write you?"

Pauline stared at him for several seconds. "She wanted to know if anything had happened to you. She was worried about you. I told her you were frightfully lazy when it came to letter writing." She paused. "Why aren't you writing her?"

Silence. "I don't know. I'm not sure we have a lot in common."

"If you want to break it off, you should tell her. You are treating her quite shabbily."

Einstein objected. "I haven't done anything."

"I think you have. From the sound of her letter I think she's heartbroken."

Einstein looked down at the floor of the carriage. "I'm sorry. I don't want to make her sad, but what can I do?"

"Write her!

"No."

Pauline fumbled with her purse, then in a low voice she asked, "Do you have another girlfriend?"

Einstein shook his head, but he thought of Mileva. He still wasn't sure what her feelings were, and he was unsure of his own feelings, but he knew he was interested."

"What do the Winteler's think of you treatment of Marie?"

Einstein frowned. "I don't know. I sometimes worry more about how they feel than I do about how Marie feels."

"They were extremely kind to you."

"I know, and I feel bad about it. But if I don't love her I can't pretend."

The carriage stopped in front of the Einstein house.

Pauline stepped down first. Einstein followed. To his relief nothing more was said about Marie over the holidays.

Einstein sat copying notes from the blackboard. On one side of him was Mileva and on the other was Grossman. The class was Dynamics and Strength of Materials and it was being taught by professor Herzog. It was the largest class Einstein was in, with approximately 140 students, most of them engineers. He leaned toward Mileva and whispered. "I won't be able to come over tonight. I forgot ... professor Stern invited me over earlier." Alfred Stern was a professor in the history of German-Jewish origins who taught at the Polytechnic.

Mileva looked at him in disgust. "I've been practicing for tonight," she whispered back. The student in front of her turned and gave them a dirty look, so they said nothing more until the end of the lecture.

"I'm sorry," said Einstein as they walked out. "I forgot."

"You do quite a bit of forgetting," she said with a look of mock scorn. The two of them had begun playing duets together, with Mileva on the piano and Einstein on the violin. They had also started to study together occasionally, but both were still cautious, neither sure where the relationship was going.

As the end of the semester approached Einstein started to get melancholy. He wouldn't see Mileva for several months and he knew he would miss her.

"Are you going home as soon as the semester ends?" Einstein asked as they walked down the sidewalk.

"I don't have anything to stay here for."

He gazed at her trying to sort out his feeling. "Would you like to go for a hike with me before you leave?" He smiled as he scratched his head. He was surprised at how nervous he was.

"I guess it wouldn't do any harm," she said teasingly.

Several days after the last day of classes Einstein and Mileva started out on one of the mountain trails near Zurich. It was a gentle trail, and would not be difficult for Mileva. This was the type of trail Einstein preferred. Many of the trails were steep, but Einstein always avoided them.

They had been on the trail about two hours when Mileva started to tire. Einstein suggested that they rest. The mountain was covered with grass and many of the spring flowers were out. There was an old gnarled tree where they stopped, but very few trees along the trail they had taken.

"Won't you get bored in Milan?" Mileva asked as she sat down.

"I'll be so busy I won't have time to get bored."

"Doing what?"

Einstein leaned back and looked up at the clouds. "My mind is bursting with ideas that I'd like to work on."

"What kind of ideas?"

"I'm particularly interested in the electromagnetic field. I feel it is not well-understood, and I'm convinced that many of the ideas scientists have about it are incorrect."

Mileva chuckled. "And you're going to change them ... set them right?"

Einstein looked down at the grass. He didn't want to appear as if he was boasting, but he had strong feelings. "I'm going to try, and somehow I feel I can." He looked up at her to see her reaction and was pleased to see her dark eyes sparkling with admiration. All at once he felt guilty. "I must be boring you; I'm talking about myself so much. Tell me about your ambitions."

Mileva leaned back as she smiled. "I would like to be

the first woman to teach mathematics and physics in a high school in my country."

"That's a worthy ambition. What about research? Are you interested in it?"

"What kind of research?"

"It wouldn't have to be experimental. I'm more interested in theoretical research."

"It sounds exciting, but I don't know if I would be good at it."

"I think you would."

As Einstein stared at her he felt he was lucky to have met her. She was someone who was his equal. She shared has interests and he was attracted to her physically.

"We could work together," said Einstein.

"That might be a problem," said Mileva.

Einstein looked at her in surprise. "Why?"

"I'm thinking of leaving next semester."

Einstein jumped up in confusion. "What do you mean – leaving?"

"I would like to broaden my education. I'm thinking of going to Heidelberg University."

"That's absurd. Why would you want to do that? Heidelberg University doesn't give credit to women."

"That's not important."

"Not important ... why would you want to go there if you don't get credit for the courses you're taking?"

"I need a change."

Einstein shook his head. "You took a semester of medicine and changed to the Polytechnic. You've only taken two semesters there. Why do you need a change so soon?"

"It will do me good."

Einstein could see it was no use arguing with her, so he dropped the subject.

The next day he was on the train back to Milan. Mileva had promised to come to Zurich in October before she left for the University of Heidelberg. He wrote her a four-page letter when he got to Milan, then concentrated on his studies.

On October 5 Mileva resigned from the physics program at ETH. The following day Einstein arrived to see her off on the train to Heidelberg.

"I still don't think this makes sense," Einstein said as he waved goodbye to her.

"Too late now."

"Write if you get bored."

Mileva smiled. "I may never get bored."

Einstein watched as the train pulled away, wondering if he would ever see her again.

In late October Heidelberg was cloaked in a thick fog and Mileva sat in her small apartment. She began a letter to Einstein "You said I shouldn't write until I was bored – and I was very obedient. I waited and waited for boredom to set in, but until today my waiting has been in vain..." She mentioned that she had told her father about him, and he had given her some tobacco to give to him. This pleased him. In his earlier letter Einstein had told her about a student who had resigned to get married. "The poor fellow probably wants to smother his love in a high romantic Swiss forest. But it serves him right; what's the point of falling in love nowadays anyway?" Einstein chuckled when he read it. At the end of the letter she described a lecture she had attended where the professor calculated the speed of an oxygen molecule.

Einstein was pleased to hear from her, but he was worried. She appeared to be enjoying herself and had said nothing about returning. He had been sure she would soon tire of the University of Heidelberg. Now he was not sure.

Chapter 6

1897 - 1899

Einstein started his second year at the Polytechnic in October, 1897. He was now familiar with his teachers, and he knew which ones he liked and which ones he disliked, and since he was not required to attend classes he skipped any that bored him. He had become close to both Besso and Grossman and had developed many casual friendships. His classmate Ehrat was Jewish as he was, and they frequently talked about the treatment of Jews around the world. But there was a difference in his life. Mileva was no longer in his classes and he missed her.

As he sat staring at the blackboard in a mathematics class he thought about Mileva. She had had a considerable effect on him and he was unsure how to handle it. It annoyed him because it was a distraction, but it was pleasant because a warm feeling came over him when he thought about her smiling face. Was he in love with her? He was sure he wasn't; he had had similar feelings for Marie and almost never thought about her now. Was it good to be thinking about Mileva so much? Not only was it a distraction, but she might not come back. It was possible that he would never see her again.

"It's a relief to finally get some physics," Einstein said to Grossman as they sat in the cafeteria sipping coffee. Einstein had been looking forward to Weber's class on physics and after the first few lectures he was pleased. "He's an excellent teacher and I'm enjoying his class." Weber's first lectures were on heat and thermodynamics.

Grossman nodded. He preferred mathematics to physics, but accepted the physics courses as part of his program, even though he was sure he would never make much use of them.

Einstein twisted the hairs of his mustache. He had started to grow a mustache, and was quite conscious of it. "I wish Minkowski and Fiedler were as good as Weber."

Grossman chuckled. "You feel that way because you like physics better than mathematics."

Einstein took a bite from his pastry. "Not necessarily. I just don't like Minkowski's style. He's so disorganized and has that curt way of talking. I think even you have to admit Weber's a much better teacher."

"I agree." said Grossman, nodding. "Weber's a better lecturer, but Minkowski an excellent mathematician. He publishes frequently."

"That may be, but do you notice that he never uses examples and never tries to make his lectures interesting."

Grossman took a sip of his coffee as he leaned back in his chair. A broad smile came over his face. "I didn't think Weber's lectures were going to be of much use to me, but they have been."

Einstein sensed a pun. He knew how Grossman felt about physics and frequently kidded him about its importance, usually to little avail." "How?" he asked curiously.

Grossman's eyes twinkled. "At one time I seriously worried when I used the toilet after somebody else. I thought I might catch something from the seat if I sat on it when it was warm. Now I know there is no danger. The warmth is just from somebody's ass."

Einstein doubled over in laughter. "And you said physics was of no use to you."

Einstein grabbed the envelope from the small table as he entered his boarding house. He could see from the writing that it was from Mileva. He ripped it open and began reading, skimming over the words, looking for an indication that she was tiring of Heidelberg. Towards the end of the letter he found it, and a sense of relief came over him. He slowed down and read each word carefully. She was thinking of coming back, but was uncertain and asked his advice. Would she be able to make up the material she missed?

Einstein smiled as he put the letter down. He had always been sure she would return; still it was a relief. He could stop worrying and thinking about it and concentrate on his studies. Knowing that she was coming back, he delayed writing. When he finally wrote he apologized for taking so long to reply. "I'm glad that you intend to return here to continue your studies. Come back soon; I'm sure you won't regret your decision. I'm convinced that you will be able to catch up rather quickly...," he wrote. He went on to tell her about the courses she had missed: Hurwitz's course on differential equations and calculus of variations, Herzog's course on dynamics, Weber's course in physics, and Fiedler's course on projective geometry.

As the end of the semester approached Einstein began looking forward to Mileva's return. He had not seen her for six months and her absence had increased his feelings for her. He tried to picture her face: her dark eyes, swept-back hair, and soft skin. He remembered the hike he had taken with her – lying on the grass, looking up at her. She was now in need of his help and he would do everything possible to help her. He sighed as he thought of Grossman; without his notes he would be of little help to her, but Grossman had always been so generous in sharing them with him, despite the fact the two were in competition with one another.

It would be difficult for Mileva, but Einstein knew she had a strong backbone. She had determination and was intelligent; he was sure that she could keep up with the lectures and learn all the material she had missed.

Einstein visited regularly with Grossman and Besso during the semester. In addition to seeing Grossman in the few lectures he attended, he also visited with him each Thursday in one of the cafes on the river. He didn't see Besso as often, but he grew continually fonder of him, and soon considered him to be a close friend. Besso wasn't as ambitious as he and Grossman, but he was a pleasant fellow to talk to, and he seemed to have a good knowledge of physics. Einstein discussed several of the problems he was working on with him and Besso always listened carefully, and occasionally surprised him with his insight. Einstein particularly liked his easy-going nature and has sense of humor. They were always anxious to tell one another the latest joke.

Einstein bumped into Besso while walking across campus one day in the late fall. It was cold and leaves from the few trees on the grounds tumbled at his feet. As usual Besso greeted him with a broad smile. Einstein always admired his fuzzy dark beard, and kidded him occasionally that he grew it to hide his face.

"I just heard from Mileva," said Einstein, stopping and touching Besso's shoulder. "She has decided to come back to the Polytechnic."

Besso raised his eyebrows. Although Einstein had said little about his interest in Mileva, Besso knew there was something between them. Einstein had introduced him to Mileva, and Besso had seen them together several times. "I'm pleased for you. I'm sure you're looking forward to her return." He put his hands in his pockets trying to keep them warm. "Did I tell you that my brother Marco is on campus this semester?"

"You might have mentioned it. I don't remember." He chuckled.

"He registered for chemistry. I'm pleased he decided to come here." Besso hesitated as he pursed his lips, then he shook his head. "I'm worried about him."

"Why?"

"He's different. He hasn't come to terms with life yet. His life is very unsettled."

Einstein chuckled. "That applies to a lot of people."

"But it's more serious in his case."

Einstein looked at his downcast eyes. It was obvious that he had strong feelings for his brother.

Einstein was relieved when the semester was over. He left for Milan in late March to send spring break with his family. Spring break was usually a time of relaxation and reading, but Einstein was tense, and the days seemed to pass slowly. He tried to study but there were many distractions. His mother was always asking him to play for her friends, and when he wasn't playing they were chatting. He took several long walks with Maja, but was glad to get back to Zurich in early April.

He stood at the platform waiting for Mileva's train. The skies were grey and a light drizzle was falling, but it didn't dampen his spirits. It had been six months since he had seen or talked to Mileva, and he had missed her. He shifted the umbrella to his other hand as he watched people get off the train. Finally he spotted her. She looked the same as he had pictured her. Dressed in a dark suit with a bow at the collar, she was looking around for him.

He felt his heart skip a beat as she spotted him and waved. He rushed up to her. "Well the little runaway has come back," he said with a smile.

"Are you glad to see me?" Mileva asked as she moved in under the umbrella. Her face was beaming.

Einstein looked her over. "The Polytechnic wasn't the same without you."

Einstein didn't feel he was on familiar-enough terms with her to kiss her so he merely gave her a hug.

"I hope this was the right thing to do," Mileva said hesitatingly. "I had second thoughts. Do you think I can catch up on all the work I missed?"

Einstein put his right arm around her as he held the umbrella in his left. They began to walk. "I'm sure you can. I'll help you," he said.

She smiled at him warmly. "Thank you."

They headed for the cover of the station. As Einstein folded up his umbrella she looked at him, then hesitatingly said, "I missed you."

Einstein took her hand and pressed it. "I missed you too."

Over the next few weeks Einstein helped Mileva organize a study program for making up the material she had missed. She took a light load in lectures so she could make up the material. Einstein was enthusiastic about helping her, but he had just come across a book on the ether and the electromagnetic field by Paul Drude that fascinated him, and he was soon spending much of his time studying it.

In late April Einstein got seriously ill and was unable to leave his room for several days. He spent most of his time in bed reading Drude. When he recovered he continued studying with and helping Mileva, but late in the semester he realized he would soon have to start studying for intermediate exams. They were given half way through the program, which would be the following October. Everyone was required to take them and Einstein knew he would have to do some serious studying for them, and he was wondering if Mileva was going to try to take them.

"Have you made up your mind about the intermediate exams?" he asked one day as they sat in the living room of her boarding house.

Mileva leaned back and folded her arms. "I don't think I should take them. I'm not ready for them. I missed so much material while I was in Heidelberg, and I haven't made it up yet. I won't be ready."

Einstein knew she was still struggling with much of the missed material, and he didn't want to push her. If she failed she might blame him, and he didn't want her to fail so soon after coming back. "I wouldn't worry," he said "you can always take it the following year."

Mileva nodded. "I think that would be better." She pushed back the book she was studying. "But it would mean I would have to take the intermediates and the finals in the same year."

"That might be tough." said Einstein. He smiled. "But I'm sure you can do it."

Einstein had missed a lot of lectures and had got behind, and with the end of the semester in sight he had little time to study. But what bothered him more than anything was that he would have to give up his extracurricular studies.

Grossman looked up from his book. He had been studying in his apartment with Einstein for several hours. During the time they had said little to one another; Einstein had spent much of the time copying and going through notes that he had missed.

"Who are you worried about the most?" asked Grossman.

Einstein cocked his head. "Minkowski, I think." The exam would be entirely oral with each teacher asking several problems which were to be solved on the blackboard in front of them.

"Why him?"

"I have less understanding of some of the things he teaches, mainly because he presents them so poorly." He paused as he looked at Grossman. "Why? Who are you dreading?

"Physics worries me more than mathematics, so I'd have to say Weber. Do you have any idea what kind of questions he'll ask?"

Einstein thought about it for several seconds. "No more than you."

"What would you do if you failed?"

Einstein chuckled. "I don't want to think about it."

Grossman took a sip from the coffee cup that was on the desk in front of him. "Are you planning on staying here over the summer to study for the exams, or are you going back to Milan?"

"What about you?"

"I think I'll stay in Zurich."

Einstein looked up at the ceiling. "I think it would be a good idea ... but I'll have to go home for a while in September. My parents would never forgive me if I didn't."

Einstein saw Mileva off on the train when the semester ended and he and Grossman spent most of the summer studying. After a short trip home in September he returned early in October for the exams.

Einstein, Grossman, Kollros and Ehrat stood in the hall waiting to be called. They joked with one another as they waited nervously. Each of them would be called before a committee consisting of Herzog, Minkowski, Fiedler, Weber and Hurwitz. Kollros was called in first.

Einstein looked over his notes as he waited; Grossman sat staring down the long hall.

"Are you worried?" Grossman asked.

Einstein laughed. "Anybody that skipped as many lectures as me should be worried. You shouldn't be. You attended them all. Tell me ... did you miss any?"

"I missed one."

Einstein shook his head, "Shame on you."

Grossman snorted. "You shouldn't be worried with all that brainpower you have."

"Are you kidding? I'll probably forget my name in there."

After what seemed like hours Einstein was called in. He could feel his heart pounding as he walked up to the blackboard. Several of his professors were seated in the first few seats in the room. Within a few minutes Einstein relaxed. He had little trouble with the first problems and began to feel confident, then one of the problems stumped him. He struggled with it for a few minutes and finally had to admit he couldn't do it. It was an embarrassing moment and shook his confidence.

"How do you think you did?" Grossman asked as they left the building together.

Einstein shook his head. "I stumbled pretty badly over

one problem, and it wasn't really that hard. What about you?"

"I'm crossing my fingers."

More orals came over the next few days, and when they ended each of the candidates was told that grades would be posted within a few days.

Mileva arrived in Zurich while Einstein was waiting for the results. He was still uncertain how well he had done, but less worried than he had been before the exams. The grades were to be posted the following morning.

Einstein and Mileva arrived at the bulletin board early. Grossman, Kollros and Ehrat were already there. They had seen their grades and were discussing the exam. Grossman spotted Einstein out of the corner of his eye and waved. He walked toward Einstein, grabbing his hand and shaking it. "Congratulations, you got the highest grade."

Einstein was stunned. He looked up at the paper posted on the board. He had got 5.7 out of 6; Grossman was second with 5.6. He smiled as he saw the result, but he was slightly embarrassed; Grossman had helped him so much and he had beat him. He turned to him. "I think I was lucky. But without your notes I'm sure I would have failed."

Mileva interrupted them. "Albert, I'm happy for you. I wish I could do that well."

Einstein turned toward her. "You will. Don't worry."

With the exams over Einstein was relieved; it meant that he could get back to the problems he had been working on before he started studying for it. He still hadn't finished Drude's book.

Just before classes began Einstein moved to a new boarding house on Kistbachstrasse. His new landlady was Frau Markwalder. The landlady's daughter Susanne, an elementary school teacher, lived with her and Einstein soon got to know her. Susanne was of medium height with short dark hair. Despite his relationship with Mileva, Einstein liked to flirt with her and tease her.

Einstein was looking forward to his third year because he would finally be taking labs. Two labs were on his curriculum: Weber's Electrotechnical lab and a more elementary lab called "Physical exercises for Beginners." Within weeks Einstein was enjoying Weber's lab and spending considerable time in it, frequently neglecting lectures as he had done previously. It was a delight to Einstein.

Einstein's second lab turned out to be considerably less delightful. Although Einstein had looked forward to it, he was soon disenchanted. The main problem was the short, stocky instructor Jean Pernet. Einstein disliked him almost from the beginning.

Pernet eyed Einstein as he walked through the door. The lab had already started and Einstein was late again. Pernet finished his announcement and began handing out the instructions for the experiment that would be done that day. He looked Einstein in the eye as he handed him the paper. His pudgy face was sullen and his eyes narrowed as Einstein took the sheet and place it on the table. He didn't like Einstein and it showed in his face, and although the feeling was mutual, Einstein ignored him.

Einstein sat down on the stool and looked at the nickel-plated calorimeter on the table and the copper cylinder. He was familiar with the function of the calorimeter and assumed he was required to determine the specific heat of the copper. The sheet giving instructions for the experiment was sitting on the table beside the calorimeter. Einstein picked it up, crumpled it and threw it in a nearby wastebasket. Pernet saw the action out of the corner of his eye; he walked quickly up to Einstein's table and confronted him.

"Why did you throw the instructions away?" Pernet asked. His face was red with rage.

Einstein shrugged nonchalantly. "I know how to use a calorimeter. I don't need instructions."

Pernet's hand was shaking. "Do you know what the object of the experiment is?"

"To determine the specific heat of the copper," Einstein said calmly.

Pernet stared at him. He had expected Einstein to hesitate, or at least give the wrong answer, and it annoyed him that he was right. "I don't like your attitude Einstein. If you do that again I'll report you to the director." He turned and walked away.

Einstein skipped the following lab, but he attended the one after it. The experiment was on thermoelectricity. This time he avoided eye-contact with Pernet as he handed out the instruction sheet. He took the sheet and put it face down on the table, then looked at the potentiometer and thermocouple that were in front of him. He inspected the thermocouple junction. Although Pernet was now at the other end of the room, Einstein was sure he was watching. He would love to have taken the instruction sheet, crumpled it and thrown it in the waste basket as he did before, but he knew that would be looking for trouble. He hooked the appropriate wires to the potentiometer, and began calibrating it.

Pernet walked quickly across the room toward his table. "Did you read the instruction?" he asked in a high, gruff voice.

Einstein looked down at the table. "I know how a

thermocouple works."

Pernet's eyes narrowed as the veins at his temples began to bulge. Without another word he turned and walked away. Several days later Einstein received a letter from the director of the Polytechnic. Pernet had reported him for "neglect of work." Einstein gritted his teeth as he put the letter down. As much as he hated to face Pernet, he was going to have to talk to him.

At the beginning of the next lab Einstein walked up to Pernet. Pernet looked at him with scorn. "Yes," he said finally.

"You sent in a report on me. I don't think it was called for," said Einstein.

"You have missed labs."

"I will make them up. Others have missed labs and reports have not been sent in on them."

Pernet stood staring at him for several seconds. Finally he said, "Einstein, you are insolent and arrogant. Furthermore, you're hopeless at physics. For your own good you should switch to medicine or law. Physics is too difficult for you."

Einstein gritted his teeth saying nothing.

"Why don't you switch?" said Pernet.

"Because, Herr Professor, I have even less talent in medicine and law. Why shouldn't I at least try my luck in physics?" He forced a smile.

Pernet snorted in disgust. He had got his revenge, but he wasn't through with him yet.

Einstein knew there was no way to get back at Pernet, and anything he did would just annoy him further. He decided to ignore him as much as possible and try not to aggravate him. He would suffer through the semester.

Several weeks later Einstein watched as Pernet argued with a student. "Why do we have to do it this way?" she asked. "It would be much easier to paraphrase the instructions in the write-up and use the same sort of table we made when doing the experiment."

Pernet's nostrils flared as his eyes narrowed. "You will do it because I told you to. It is the required way to write up experiments at this institution."

Einstein walked up to her after Pernet had left. "He's quite a clod, isn't he?"

"He sure is. I can't stand him." She looked more closely at Einstein. "I know who you are. You're Mileva's boyfriend."

Einstein smiled.

"I'm Margarete von Uexkull. I board in the same house Mileva does." She shook her head. "I guess I'll just have to do what he says, but it's confusing. This lab is hard for me."

"Why don't you give me your data. I'll write it up for you."

Margarete looked at him in surprise. "I couldn't do that."

"Why not?"

She thought about it. "Would you? I'd appreciate it."

Einstein told her he would. Two days later he gave her the written-up lab. A week later Einstein saw her in the lab. "Thank you," she said. "I got a good grade for that lab and guess what professor Pernet said." She mocked his high-pitched voice as she repeated his words, "See I told you that you could do a good job if you set your mind to it." She laughed as Einstein smiled.

It was a dark, moonless November night. Einstein sat at his desk engrossed in a book. A light knock came on his door. He got up and walked to it slowly, wondering who would be visiting him so late at night. Opening the door he saw Besso standing before it; his eyes were red as if he had been crying. Einstein knew something tragic had happened.

"Michele my friend, come in," he said. "What's wrong?"

Besso rubbed his eyes.

"Here, have a seat," said Einstein.

Besso sat down, then looked up at Einstein. "I told you about my brother Marco."

"Yes," nodded Einstein. "He was majoring in chemistry."

"He committed suicide," said Besso.

A look of alarm came over Einstein's face. "He's dead?"

"Yes."

Einstein put his hand on Bess's shoulder. "I'm sorry."

"I knew something was going to happen," said Besso. "He was so discontented, but I never thought he would do this."

Einstein sat down beside him. He knew there were few things he could say that would comfort Besso, but he wanted to help him as much as he could. "I know how close you were to him." he said.

Besso leaned over and put his head between his hands. "I tried," he said hesitatingly. "I did my best."

"I'm sure you did," said Einstein with feeling. He knew that Marco had been distraught for some time, and he was tempted to say, "maybe it was for the best." but he held his tongue.

Besso looked up at him, his red eyes shining in the light from the gas lamp. "Everything I did ... it didn't help."

Einstein moved his head slowly from side to side. "I know how you feel." He had worried about his father after the two business failures, but he had yet to see a serious personal tragedy. Still, deep down, he was sure he knew how Besso felt. He continued talking to him and comforting him until well after midnight. Besso was embarrassed when he realized how late he had stayed, and he thanked Einstein several times at the door as he left.

Soon after moving into his new boarding house Einstein learned that Susanne Markwalder played the piano, and within a short time he was playing duets with her.

Susanne sat on the piano stool watching Einstein put his violin in the case. They had been practicing a duet. "What's your girlfriend's name?" she asked. Einstein had casually mentioned that he had a girlfriend.

"Mileva."

Susanne twisted her mouth as she rotated the piano stool back and forth. "What's she like?"

Einstein closed the case and looked at her with a smile. "She's very clever."

"What does she look like?" Susanne was much more interested in her appearance than her intelligence.

Einstein chuckled. "Why are you so curious?"

Susanne shrugged. "I guess I'm just curious by nature." She looked at the music on the piano, trying to avoid his eyes. She didn't want to appear as if she was seriously interested in him.

Einstein started to leave the room, then stopped and looked back at her. "You'll see her tomorrow night. I've asked her to come over."

The following evening Susanne watched from behind a cracked door as Einstein walked down the steps and opened the front door. He took Mileva into the living room rather than to his room upstairs, knowing that there would soon be considerable gossip if he didn't. Susanne watched as they talked and studied. Mileva seemed very reserved to her, with Einstein doing most of the talking and laughing.

The following morning as she stood by the sink her mother asked, "Was your curiosity satisfied?"

Susanne shrugged.

"What was she like?"

"She's not much of a match for Albert. I think she's very plain."

"Maybe she has a pleasing personality. There are other things besides beauty."

"That may be, but she seemed very quiet to me. Personally I don't see what he sees in her."

"I think maybe there's a touch of jealousy there," her mother teased.

"You think I'm interested in him?"

"You're spending a lot of time with him."

"It's just that I enjoy his company."

Einstein was sure that Susanne had seen Mileva and later that day when he encountered her in the hall he asked, "What did you think?"

"Think about what?" she said innocently, as if she didn't know what he was talking about.

"Mileva, my girlfriend."

"Oh," said Susanne. "She seemed very pleasant." She knew she couldn't tell him the truth.

Einstein was pleased. "She's a very intelligent girl," he said proudly.

"It's a good thing she's intelligent," Susanne said to herself. Then out loud, with a smile, "Would you like to practice Schubert's Sonata tonight?"

"I'd love to," said Einstein.

Susanne hesitated. "One of the boarders asked if she could sit in the next time we played." She watched his face for a reaction. "She may bring her two daughters. Is that all right with you?"

Einstein nodded. He always enjoyed playing for an audience.

That evening the three women sat on the sofa as Einstein and Susanne began playing. Soon all three were knitting as they listened. At first Einstein didn't notice the clicking and banging of their needles, but he watched them closely as he played. They were paying more attention to their knitting than the music. Then he and Susanne came to a pianissimo section and the clicking and banging of their needles resounded through the music. At almost this moment one of them dropped a stitch and sighed loudly. The others laughed softly.

Einstein's agitation increased. Finally he could stand it no longer. He stopped playing; Sussanne also stopped, wondering what was wrong. She waited for him to say something. Without a word, however, he put his violin in his case.

"Why are you stopping?" asked the mother of the two girls.

Einstein looked at her in disgust. "Because we would not dream of interrupting your work." He picked up his case and walked out of the room.

Susanne was still sitting at the piano. Her mouth dropped as she watched Einstein walk up the stairs to his room. Later that day he apologized to her.

It was snowing heavily, with large flakes floating down out of a dense, grey sky. Every so often the wind would swirl the falling snow, making it into shimmering white curtains. Besso had on a heavy coat and high boots. Snow covered his beard. He stomped his feet as he walked into the library. To his delight he saw Einstein sitting at a long oak table with several books in front of him. Sitting down beside him he said, "Greetings, I haven't seen you for a while." He undid the top button of his coat.

"I've been working on some new ideas," said Einstein.

Besso glanced at the books sitting on the table. "I've come across several books I think you would be interested in. Have you heard of Ernst Mach?"

Einstein shook his head. "Not that I can remember ... but you know my memory." He laughed.

"He has some interesting ideas. Come ... I'll show you." He got up and started toward the shelves. Einstein followed him.

The books were on an upper shelf and Besso had to stretch to reach them. He pulled down Science of Mechanics and handed it to Einstein. Einstein looked briefly at the table of contents, then flipped through the book. "I'll check it out," he said.

Over the next few days Einstein read the book and was impressed. Mach suggested that the concept of absolute space and absolute motion be re-examined; he was suspicious of the way things were currently treated by scientists. He criticized many of the accepted concepts of physics, even going as far as suggesting that Newton's laws of motion may be wrong. Einstein admired his brashness. He saw in him a new breed of scientist – one that was able to see the world as a child sees it – awe-inspiring and full of wonder.

On his next visit to Mileva he told her about Mach's books.

"I can tell by your eyes that his ideas fascinate you," she

said.

Einstein smiled. "They may be what I need in my reformulation of electrodynamics. I think his approach is the right one."

"Maybe we can study the book together,"

"That would be a good idea," said Einstein, looking closely at her dark smiling eyes. He was still delighted that he had found her. They had so many interests in common, and she was everything he wanted in a woman. He thought of how much joy she had brought into his life. He had to admit that she was gloomy at times, but he overlooked this, sure that it was just a passing phase. "Do you have a photograph of yourself?" he asked.

Mileva looked at him in surprise. He had said little of his feelings for her, but she knew they were growing stronger every day, and she knew she was in love with him. "Why?" she asked, leaning back in her chair and grinning.

"I'll be going back to Milan for spring break and I would like to show my family a photo of you."

She frowned. "Maybe they won't like me if they see my photo."

Einstein laughed. "That's unlikely. I think they'll love you." He knew that both of his parents looked down on the darker races from Northern Europe, but he wasn't worried. He had already planned what he would say.

Mileva left the room; after digging around in a bureau for a few minutes she returned and handed him a photograph. "Is this one all right?"

Einstein spent several seconds looking at it. Finally he said, "It's lovely, but it hardly does you justice."

"You're a darling," said Mileva moving closer to him on the sofa.

Einstein put his arm around her, conscious of her soft

skin and perfume. He looked at her throat, her firm chin, then into her dark shining eyes. Mileva gazed back at him calmly, knowing that she loved him, hoping that he felt the same way about her. They said nothing for several minutes, then the room was filled with a silent tension. Einstein moved closer to her and brought his lips up to hers.

Einstein left in late March for Milan. He would have almost two weeks with his family. His mother picked him up at the station. She was very quiet during the ride to their house. Einstein had been worried about his father's new business, but little had been said about it, so he assumed everything was going smoothly. Despite the setback they still lived in a large, well-furnished apartment.

Maja got home from school shortly after Einstein arrived, and as usual she was delighted to see him, throwing her arms around him and hugging him. Einstein talked briefly to Maja, then turned to his mother. "Mamma, I have a new friend at the Polytechnic."

Pauline stared at him briefly. "A girlfriend?"

"Yes," nodded Einstein. "I have a photo of her." He pulled out his wallet, took out the photograph and handed it to her.

Pauline studied it carefully. "She looks dark. What nationality is she?"

Einstein was reluctant to tell her that Mileva was Serbian, as he knew how she felt about Serbs. "She's from Southern Hungary."

"How old is she?"

"Twenty-one," said Einstein cautiously.

Pauline looked at him in surprise. "She's four years older than you?"

"Three and a half."

"That's too much."

Einstein watched as his mother looked again at the photo. "She's a very clever girl," he said.

Pauline handed the photograph back to him without saying anything more.

"Let me see it," said Maja. Einstein handed it to her.

"She looks nice," said Maja after looking at it briefly. "But she's not as pretty as Marie."

"She has a beautiful singing voice," said Einstein.

"Have you kissed her yet?" asked Maja with a giggle.

Einstein had gone through this with Marie and he expected it. Maja loved to tease him, and in many ways it pleased him. "None of your business," he replied playfully.

Maja continued to tease him during his visit home, and occasionally Pauline joined in, but Einstein was worried about his mother's reaction; he was sure she had reservations and was saying little.

He was met by Mileva at the train station in Zurich upon his return. Within moments he had told her that he showed her photo to his family.

"What did they say?" asked Mileva with a worried look on her face.

"They loved you."

"Did they really," said Mileva as though relieved.

"Oh Johnnie, I'm so happy," she said kissing him. "I'm going to call you Johnnie from now on."

"That means I have to have a name for you." Einstein brought his hand to his head pretending to be in deep thought. "Let's see ... Dollie. That will be a good name. We'll be Johnnie and Dollie."

Mileva laughed.

Einstein worked on an experiment in Weber's lab.

He liked Weber's lab much better than Pernet's. It was extremely well-equipped and he had done several experiments in addition to the ones that were required. Over the semester he had spent a considerable amount of time in the lab, and while he had been reading about the ether and electromagnetic waves an idea had come to him. He was sure he could test it in the lab,

Einstein waited until late in the day. Weber was busy grading labs as Einstein approached him.

"I would like to ask your permission to do an experiment." Einstein said.

Weber looked up at him. Early on he had been sure that Einstein would be an excellent student, but lately he was beginning to have reservations. "What kind of an experiment?"

"I'm interested in the ether and believe that I know a way that it can be tested."

"What is the object of the experiment?" asked Weber skeptically.

"I would like to measure the velocity of the earth relative to the ether."

Weber took a deep breath and sighed. "How?"

Einstein went over the experiment in his head. "I would need a thermocouple and a semi-silvered mirror that allowed some of the light to get through, and some to be reflected. If the mirror were set up so that some of the light from a source passed through parallel to the motion of the Earth, and some was reflected in a perpendicular direction or possibly an antiparallel direction there would be an energy difference between them because of their motion relative to the Earth, I could measure this difference using the two thermocouples."

Weber sat staring at him without saying anything. "And

you want to perform this experiment?"

"Yes."

"I don't think it is possible. The energy difference would be too small to measure, even with thermocouples."

Einstein pursed his lips. He hated to contradict Weber, but he was certain the experiment would work. "We don't know for sure that is true until we do the experiment."

Weber looked down at the books on his desk, "You have been skipping a lot of lectures lately."

Einstein was unsure how to reply. Grossman was giving him all the notes he missed, but he didn't want to say anything about this to Weber, so he said nothing.

After several minutes of silence Weber finally said, "I'll have to take this under advisement. I'll discuss it with some of the other faculty and let you know."

"When?" asked Einstein.

Weber raised his eyes toward the ceiling. "Check with me on Monday."

Einstein was sure the experiment would work, and he was pleased with his ingenuity, sure that the results would be of great significance. He could hardly wait.

The following Monday Einstein reported to Weber's office. Weber came to the point immediately. "I've discussed your experiment with some of the other faculty," he said curtly. "We've decided that it can't be done, and would be a waste of time and resources."

Einstein's jaw dropped. He began to protest, but thought better of it. He knew that Weber's mind was made up and anything he said would not help. He nodded as he gritted his teeth and left without saying anything.

"Damn," he muttered to himself as he walked to Pernet's laboratory. He was late for the lab that day but he didn't care. All he could think about was Weber's stupidity.

It was such an important experiment, and it should be performed.

Pernet had already begun as Einstein walked in and seated himself.

"We will be titrating the silver from silver nitrate today," said Pernet. "The residue, when dry, can be explosive so it is imperative that you follow the written instructions very carefully."

Einstein stared off into space without hearing a word. He was still annoyed. When Pernet finished his brief talk Einstein looked at the apparatus on the table. Beside it was the written instructions. He glanced at them, then shoved them to the side. He was sure he knew how to do the experiment. Pernet had stopped coming over to his table and Einstein was glad; he hated him peering over his shoulder as he performed experiments.

Einstein looked over at Ehrat, who was at the next table. Ehrat was majoring in mathematics, but he had taken the lab as an option. Einstein wondered why. As much as he liked working in the lab, he hated Pernet's lab. But more than anything he hated Pernet's attitude and air of authority.

Einstein went through the steps of the experiment mechanically with a feeling of indifference. He hadn't looked at the instructions, or listened to Pernet and was unaware of the danger. It was late in the day and he was finishing up. The nitrate residue had dried in the aluminum container on the desk. As Einstein went to move it an ear-shattering explosion echoed throughout the lab. Einstein had no idea what had happened. He pulled his hands back; they were black and bleeding.

Most of the other students in the class stared at him with widened eyes but nobody moved. Finally Ehrat ran over to him as Pernet rushed from the opposite end of the room.

"I told you to be careful," Pernet yelled at him.

Einstein held his blackened hands up. Blood was now pouring from one of them.

"You'll have to go to the infirmary," said Pernet. "Have them clean you up and check your hands. You may need some stitches."

"I'll go with him," said Ehrat.

As Einstein exited the room, Pernet's upper lip curled and he smiled. "Serves him right," he muttered to himself.

Chapter 7

1899 - 1900

Einstein sat on the edge of his bed unwrapping the bandages on his hand. It had been over a week since the explosion, and he was still worried about how the wound would affect his violin playing. The violin was an important part of his life; he couldn't imagine life without it. He thought back to the accident as he clipped the last bandages. He knew he had no one to blame but himself. On the way to the infirmary Ehrat had remind him of Pernet's warning and the warning in the instructions. He was embarrassed to tell him that he had not heard or read about the danger.

As the last bandages came off he looked at the wound on his hand. It was not as bad as he had expected, and it was healing well. He moved his fingers back and forth, sure now – and relieved – that it would have no serious effect on his violin playing. The wound continued to heal, and three weeks later there was little sign of it.

With the end of the semester in sight Einstein was looking forward to the summer break. He had discovered several new books and was anxious to read them. A new book with new insights was always a joy to him; he looked forward to each one with a feeling of excitement. His quest for understanding was being met mostly by self-study. Lectures were important, but much more stressful and less enjoyable than delving into books and getting the answers on your own. He hated the memorization required in lectures; to him, understanding was the most important thing.

"I'm not looking forward to the summer," said Mileva as she stood waiting for her train. "My brain is going to be aching by the time I get back here in October." She would be spending the summer preparing for the intermediate exams.

"It will be good for you," Einstein said with a laugh. "It will keep you out of mischief."

"What kind of mischief could I get into?" asked Mileva, twitching her nose. She grabbed Einstein's hand and pulled him toward her.

"A pretty girl like you ... you can never tell," he teased.

"And what will you be doing?" Mileva shot back.

"I'll have my nose buried in books all summer."

"I'll bet." She gave him a hug. "You better write me every day."

Einstein mouth dropped as his eyes widened. "Every day," he repeated in mock astonishment. "That may be difficult, but I will write."

The train whistle signaled for everyone to board. Einstein gave Mileva a quick kiss.

"I'll miss you," she said as she began boarding the train.

"And I'll miss you," Einstein said, letting go of her hand. He smiled broadly as he said good-bye.

The following day Einstein boarded a train for the short trip to Mettmenstetten; it was a small resort between Zurich and Zug. His mother and Maja were staying at the Hotel Pension Paradise. The train took him along Zurich Lake. As he looked down on the lake he thought of the problems he was interested in, and the books he planned on reading. He had books by Mach, Boltzmann, Kirchhoff, Helmholtz and

Hertz with him, more than enough to keep him busy over the three months he would have off.

His mother and Maja were at the station in Mettmenstetten. The air was warm and clear as he got off the train, the landscape brilliantly green, and lush with wildflowers and grass, and the freshness of late spring. The sun was high in the sky, a sky that was as blue as the surface of Lake Zurich. Maja came running when she saw him, and was soon clinging to him in an intense hug. She had grown and matured, and Einstein saw that she was wearing a small amount of makeup. At seventeen, she still resembled Einstein, with the same dark wavy hair, high forehead and dark eyes. Pauline was directly behind her. "When do I get my hug?" she said as Maja continued to cling to Einstein.

"Hello Mamma," said Einstein as he released Maja and gave her a hug.

The Paradise Hotel was only a short distance from the station, and Einstein would have preferred to walk, but his mother had a carriage so they rode. He was anxious to try some of the nearby trails, and Lake Zug was not far away. It was considerably smaller than Zurich Lake but beautiful in its own right, and there were many hiking trails around it.

Einstein soon settled into a routine, studying and reading the books he had brought in the morning and hiking, playing the violin and generally enjoying himself in the afternoons. But there were distractions and he soon began to resent them. His mother seemed to know every older woman in the hotel, and there were frequent visits. Their chattering got on his nerves, and made it difficult for him to study; furthermore, his mother frequently asked him to play for them.

Einstein was sitting on the back steps of the hotel when the sister-in-law of the owner, Robert Markstaller, approached him. She had dark-brown hair and unusually large eyelashes that, strangely, were much lighter than her hair. She was seventeen, and very conscious of every boy her age that she met. Einstein had talked to her several times, flirting with her as he did with all young girls. He had even teased her about not having a boyfriend. Her name was Anna Schmidt.

"Can I ask you a favor?" Anna said as she sat down on the grass before him. Her face was slightly flushed, but her eyes shone with interest.

"You can ask," said Einstein. "There's never any harm in asking." He loved to tease, and this morning he was in a good mood.

"You're sure."

Einstein ran his hand through his thick black hair. "I'm at your service ... your faithful servant. What will you have me do?" His eyes sparkled as he grinned.

"Will you sign my autograph book?" She had been holding it behind her back. Bringing it out she thrust if toward him.

Einstein looked at it and began thumbing through it. "It looks like you've got some pretty important autographs here," he said teasingly. He saw that most people had written short verses along with their autograph.

"Do you want a little verse along with it?"

Anna's eyes lit up. "Please."

Einstein looked up at the sky as he thought, then he wrote:

Little Girl, small and fine
What should I inscribe for you here?
I could think of many things
Including a kiss
On your tiny little mouth.

If you're angry about it
Do not start to cry.
The best punishment is –
To give me one too
This little greeting
Is in remembrance of your rascally little friend,
Albert Einstein

Einstein handed the book back to her. As Anna read through what he had written she started to giggle, then she looked up at him with widened eyes. "You wouldn't really want to kiss me, would you?"

Einstein laughed. "You never can tell."

"I'd have to tell your girlfriend."

"You wouldn't," he said, faking surprise. "That would never do. It would have to be our secret."

Einstein enjoyed the quiet mornings of study and contemplation. They were a relief after the weeks of cramming he had just gone through. There was no pressure on him now; still, he knew he had no time to waste. There was too much to be learned. Nevertheless, he took many afternoons and evenings off to go hiking. Shortly after he arrived he went on a trail near Zug Lake with Robert Markstaller. Lake Zug was only eight miles long, compared to Lake Zurich's twenty-five miles, but it had a charm all its own. It was enclosed by gently sloping wooded hills that frequently gave a view of the majestic snow-peaked mountains behind. And towards the end of the summer he traveled to Mount Santis with Maja, hiking on the same trail he had several years earlier when he had almost fallen to his death.

Despite the sympathy and compassion he felt for his

mother and Maja he eventually became disenchanted with them. Gone was the laughter of earlier days; they seemed so serious, so preoccupied with insignificant things. They were so different from Mileva. He could talk to Mileva about physics, about his hopes and discoveries in away he couldn't with them. They were becoming boring.

He sat at his small desk in his room. "Here in Paradise I live a nice, quiet philistine life with my mother hen and sister," he wrote. He told her about the books he was reading and the ideas he was developing, then he thought of her plight, and how hard she must have been studying. "You, poor girl," he wrote, "must be stuffing your head with gray matter, but I know that with your divine compassion, you'll accomplish everything with a level head."

He always felt better when he finished a letter to her; he knew she would enjoy getting it as much as he enjoyed getting a letter from her. But he chuckled to himself as he walked to mail it. She would be "pouting and fretting" if she knew what a good time he was having hiking, loafing and studying what he wanted to study. He knew her nature. She could be a "little witch" at times, but he loved her and wished she could be here with him.

As he approached the hotel he could hear his mother talking with one of her friends. The constant chatter usually bored him, but he stopped now as he heard the words, "She's Serbian and walks with a limp." His fingers clenched as the hair in the back of his neck bristled. He continued listening, but no more was said about the "Serbian." He finally moved off.

"What did the remark mean?" he asked himself. It was the tone of her voice that bothered him more than anything. Did she really dislike Mileva? He wondered if he should tell Mileva about it, knowing that it would worry her. But it was something he couldn't keep to himself. In his next letter he mentioned it, trying to soften the blow by adding, "maybe I just imagined it."

Einstein looked forward to the mail each day. He was always delighted if he got a letter from Mileva, but he also got other letters. He was quite surprised when he got one from Julie Niggli of Aarau. He had written to her several times since he had left Aarau, but hadn't heard from her in quite a while. "What kind of mischief is she getting herself into now," he muttered with a chuckle as he opened her letter.

His eyes widened as he began reading. She was in love and her suitor had asked her to marry him. Einstein laughed to himself. "Poor girl." She asked his advice. He thought about it, then realized he would like to deliver it in person. A break from the constant chatter of his mother's friends would be a welcome relief. He had played duets with Julie when he lived in Aarau and would like to do it again. But he couldn't tell his mother he was going to Aarau to visit Julie; he needed a better excuse. Then he remembered that the Wintelers had a visiting scientist, Dr. Haab, staying with them; he would tell his mother that he had to talk to Dr. Haab. He wrote a short letter to Julie, telling her he was coming and cautioning her against her suitor, and he also wrote the Wintelers.

His main worry in going to Aarau was the chance that he might bump into Marie. What would he do if he met her accidentally? What would he say? How would she react? He was sure he was immune to her charm now, but he didn't want to take a chance of bumping into her, so he asked Julie to inquire. She wrote back that Marie had been there recently and was not likely to return for a while. Einstein was relieved.

Pauline Winteler met him at the station. He was delighted to see her. Her hair was starting to gray and there were small wrinkles at the end of her eyes, but she was the same warmhearted person she had been when he lived there, and she was still like a mother to him, yet so different from his own mother.

"How is papa?" Einstein asked when they were in the carriage, referring to Jost Winteler.

"Busy as usual. And he is still talking to the birds." She laughed heartily.

Einstein then asked her about the children, but was careful to avoid asking about Marie. He knew he wasn't fooling Pauline, but he also knew that she wouldn't embarrass him.

"I think I've talked Maja into coming to school here. She is seventeen now, and could register next fall."

"Albert, I'm pleased," she said, squeezing his hand. "I hope you told her she can stay with us."

Einstein's face beamed. "She knows how much I enjoyed it. I'm glad you'll have her. I'll tell her."

Julie Niggli was at the Wintelers when the coach arrived. She was anxious to get Einstein alone. His letter had disturbed her and she had to talk to him about it. He was unpacking his suitcase when she knocked lightly on his door.

"Come in," he said as he continued unpacking. When he saw who it was he smiled.

"Albert, I've been wanting to talk to you."

He knew what she wanted to talk about. He motioned for her to take a seat in a nearby chair as he sat down on the bed.

She sat down. "I'm worried about what you said about my fiancé." She shrugged as she pulled a face. "I can't believe that I won't find happiness with him. I'm in love with him."

Einstein grinned. "He's a man, and I know what they're like. As I said he'll be sullen one day, high spirited the next, then cold and depressed on the third."

"But he seems so nice. He's such a gentleman."

"That's because you don't really know him yet. Men are men. If you marry him I can assure you he'll be selfish, ungrateful and he may even be unfaithful."

"Are you serious?" she asked, her eyes large and intense, her face showing some fright.

Einstein thought about his treatment of Marie and he knew that much of what he had said applied to him. Maybe it was because he considered her intellectually inferior. He wondered if he would eventually treat Mileva the same way. There was, of course, a difference: Mileva was his intellectual equal, or at least nearly so. "I'm afraid so," he said. But he was not interested in talking about her love life; he had come to Aarau to play duets with her "Did you bring some music?" he asked.

"Yes, but you've got me so worried. I don't know if I will be able to concentrate on it."

"You should come to Paradise sometime. We could make beautiful music together."

Julie looked at him in shock. "The Paradise Hotel ... with you?"

Einstein laughed. "We'll be well chaperoned. My mother and sister are there."

Einstein and Julie played several duets together that evening, with most of the Winteler family watching and listening. But Julie kept asking him for advice and it eventually began to annoy him.

To his surprise and delight he got a letter from Mileva

that had been forwarded from Mettmenstetten. She was "trapped" on her farm at Kac; an outbreak of scarlet fever and diphtheria in the nearby village of Novi Sad force her and her family to stay out of the town. She reported that her "cramming" was proceeding slowly. "Fiedler is my biggest headache and the material is the hardest to master." She then asked him for advice about the exam. "Do Fiedler and Herzog ask specific things, examples, or do they ask only general questions?"

Her letters were starting to get gloomy and it bothered Einstein. He preferred perky, uplifting letters like the ones had had got from her at Heidelberg. "Don't you feel sorry for me?" she wrote. And it was obvious that she was beginning to worry about Pauline. "I hope you're not letting anyone read my letters; you must give me your word."

Einstein put the letter down as he thought about her. She was obviously much more worried about the intermediate exams than he had been; furthermore, she was studying by herself and would be taking the exam in October alone. He had had Grossman. He did feel sorry for her.

He talked several times to Professor Haab over the next few days, and played duets in the evening with Julie, then he left for the Paradise Hotel.

About the middle of September, Einstein, his mother and Maja returned to Milan. Einstein was continuing to study books by Helmholtz, Boltzmann and Mach. He continued to reassure Mileva in his letters to her. She was still seriously worried about the intermediate exams. "You shouldn't let this little exam bother you too much. That should be easy for you -- especially with such harmless competition," he wrote.

Maja had now decided to go to the Cantonal school in

Aarau as Einstein had, and she was excited about it.

"Will you go to Aarau with me?" she asked.

Einstein frowned. He had been at Aarau only a month earlier, but always enjoyed going there. There was, however, the usual problem. "What if I bump into Marie?"

"Why are you so worried about that? Are you embarrassed about the way you treated her?"

"I didn't treat her that badly," Einstein objected, "We just decided to break up."

Maja looked down as if she didn't believe him, but she dropped the subject. "I'd appreciate it if you would go with me. You don't have to stay. Just drop me off."

"That sounds okay to me," said Einstein with a smile.

Two days later he accompanied Maja to Aarau. He stayed only an hour, talking briefly to Pauline Winteler. Maja inquired about the children and was told that Marie would be coming the following day. Einstein was relieved that he had missed her.

On the train back to Zurich, Einstein thought about Mileva. Her exams would be over now, and he hoped she would be in a better mood. He was sure she would pass, even though she was seriously worried. It would be as much of a relief to him as it would to her. She met the train as it pulled into the station.

"You little rascal. It has been days since you finished your exams, and you never wrote me and told me how you did," said Einstein as she ran up to him.

She hugged him for several seconds without saying anything.

"Well, how was it?" he asked again.

Mileva took a deep breath and sighed. "It was horrible. Just as bad as I thought it would be." She paused. "Fiedler intimidated me. I could hardly talk."

"How did you do?"

She frowned. "I passed, but the competition wasn't as harmless as you said it would be. Six of us took the exam and I ranked fifth."

"That's all right," said Einstein. "The important thing is that you passed."

"I'm glad they're over," said Mileva with a sigh.

"You can relax now."

In early November, 1899, Einstein moved to Unionstrasse so he would be closer to Mileva who was on Platterstrasse. This was the fourth and final year of their program, and they continued studying together as much as possible.

Einstein sat at a desk in the living room of Mileva's boarding house. He sighed as he leaned back and rested the book on his lap. "I'm more impressed with Helmholtz every time I read him, and I agree with him. He suggests that every theory not verifiable by observation or experiment be rejected. To me this applies to the "ether." Its reality has never been verified, and until it is I cannot accept it. I also like his ideas on the relationship between the ether and matter. Too bad he died before he could test them."

"You study too hard," said Mileva.

Einstein shook his head. "Some big changes are coming. I can feel it in my bones. If I delay too long others will made the discoveries." He nodded towards Helmholtz's book.

"Come over here," Mileva said who was sitting on the couch. He put his arm around her as he sat down.

"Let's talk about us," said Mileva.

Einstein leaned over and kissed her on the cheek. "What about us? We're going to get married aren't we?"

Mileva sat up quickly as a smile came over her face. "You haven't asked me." She looked him in the face, studying his eyes.

"I'm asking you now."

Mileva grabbed him. "Oh Johnnie, I was hoping so much that you would ask. I'm so happy."

"So what's your answer?" Einstein asked confidently.

"You know what it is."

"But I want to hear it."

Mileva kissed him softly, then leaning back said, "Yes." She smiled broadly, then a frown came over her face. "What about your mother? I know she doesn't like me."

"We don't know that for sure. She wrote 'Greetings from the family' on the last letter I wrote to you. Besides, I can handle her. Don't worry."

Mileva continued frowning. "I am worried." And she continued to worry over the next few weeks.

Einstein could not make himself believe that his mother disliked Mileva. He knew there was a problem, but he continued to dismiss it. When he went home at Christmas, however, for the first time he began to get seriously worried. Pauline said nothing about Mileva; she never asked about her, never inquired about her health and she never mentioned her name. To Einstein it was unnerving. He finally asked Maja about it.

"Mamma has never said anything to me about her," Maja replied when Einstein asked.

"Does she dislike her?" asked Einstein wrinkling his forehead.

Maja said nothing for several seconds, then sighed. "I don't think she likes her. She talks about her to her friends occasionally and doesn't have anything nice to say about her."

Einstein shook his head as he looked down at the floor. "How could she dislike her. She's never met her." He was annoyed.

"She dislikes what she knows about her."

Einstein stared at Maja briefly. "I've proposed to her," he said.

A look of surprise mixed with joy came over Maja's face. "You're going to marry her?"

"Yes."

"Oh, Albert, I'm so happy for you." She paused. "But what will Mamma say?"

"That's what I'm worried about."

Einstein was relieved to leave Milan. The atmosphere seemed stiff and tense to him, and he was sure some of the problem centered on Mileva. As much as he wanted to get it out in the open, he said nothing to his mother about it.

For several years Einstein had been saving 20 francs each month toward the money he needed for citizenship. He now sat before an examining board of eight members in the first stage of the citizenship process.

"Do you drink spirits?" one of the examiners asked.

"No," replied Einstein.

"Do you lead a respectable life?"

"Yes."

"Are you syphilitic?"

"No."

"Was your father or grandfather syphilitic?"

"No."

Many of the questions embarrassed Einstein, but he answered them all truthfully. After about an hour the questioning stopped, and he was asked to leave the room while they discussed the question of citizenship. It looked bleak for him. He was of little means and the board wanted to be sure that new citizens never became wards of the state.

A few minutes later one of the men came out and told him that everything appeared to be satisfactory, but the process was not yet over. A detective would be checking out his story over the next few weeks and he would be notified.

He was relieved as he left the building, sure that within a short time he would be a Swiss citizen. He told Mileva about it that evening, but she was still gloomy with worry about his mother's feelings, and she said nothing in reply.

A few moments later she muttered, "Helene Kaufler is going to Milan." Kaufler was a close friend of hers, a history student at Zurich University. "She has never met your parents and would like to go and see them."

Einstein stared at her with a puzzled look on his face. "My parents don't know that she is a friend of yours, do they?"

"No, they don't."

"We could get her to ask Mamma about you. That way we'd know for sure. Do you think she would do that for us?"

"I'm sure she would if I asked her."

The following evening Mileva and Albert talked to Helene, and she agreed to visit his parents and ask about Mileva.

Albert and Mileva were anxious to see Helene when she returned several days later. They met with her the evening she returned.

"What did she say?" Mileva asked anxiously as she took Helene's coat.

Helene took a seat on the couch. She looked toward Einstein. "Your mother was very nice to me. I enjoyed her company and I also met your father. He was very distinguished-looking and handsome."

Einstein beamed with delight.

Helene then turned to Mileva. "We talked about you". She hesitated as she continued. "She doesn't have a very good opinion of you, I'm afraid to say. I hate to have to tell you, but you told me to tell the truth even if it hurt. She feels you're not good enough for her Albert. She's disappointed that he's involved with someone who is so much older than him, and she puts most of the blame on you."

"She doesn't like me at all?" Mileva asked in a frightened voice.

Helene looked at her in silence. "She ridiculed you." She paused as she looked down at the floor. "I have to admit I went along with her at times so she would tell me more. But none of it was good."

Mileva sat staring at Helene in shock.

"I'm sorry I had to tell you this," said Helene, trying to console her.

"So you thought my old man was handsome," said Einstein, trying to change the subject.

Helene looked at Mileva in sympathy, then turned to Einstein and nodded. A few minutes later she told them she had to leave. They both thanked her at the door. She had barely got out of the door when Mileva turned to Einstein. "She hates me. Oh Johnnie, what can we do? We can't get married."

Einstein put his arm around her. "You're taking it too hard. She doesn't hate you, and we're still going to get married. I told you earlier I'd look after everything and I will. Don't worry about it."

With the end of the semester only days away, Einstein copied into his notebook as Weber finished up his lecture.

"This is as far as I'm going," said Weber as he put the

piece of chalk down. "I will spend some time reviewing over the last few lectures, so if you have any questions bring them to me then."

Einstein sat in stunned silence. He had been waiting all semester for Weber to cover Maxwell's equations. He couldn't believe what he was hearing. He waited until the classroom was empty, then he walked up to Weber, who was now cleaning the blackboard. "Herr Weber," he said.

Weber turned and looked at him with an emotionless, almost sullen face.

"Are you not going to cover Maxwell's theory?" Einstein asked.

"This is as far as I'm going. Maxwell's ideas are very speculative and I don't believe they belong in the classroom. If you wish to learn about them, you'll have to study them on your own."

Einstein gritted his teeth. "Maxwell's theory is an important part of electricity and magnetism. I don't think it should be skipped."

"That may be your opinion, but it is not mine, and I'm not going to cover it," said Weber curtly. He turned and continued erasing the blackboard.

Einstein stood looking at the back of his head for several seconds, then walked out of the room.

Later that day he sat in the Cafe Metropol with Grossman. "I can't believe he would skip such an important part of physics," said Einstein, looking grimly down at the table.

"I don't mind," Grossman joked. "You've studied it on your own anyway, haven't you?"

Einstein nodded. "Yes, but I was still looking forward to taking it up in class." He sipped his coffee and took a bite from a wiener. "And Minkowski ... his class is titled "mathematical physics – the mechanics of rigid body motion" and so far all I've seen is pure mathematics. He didn't even define what a rigid body is, and he hasn't discussed the concepts of force, impulse or momentum. How can he do mathematical physics."

"Complain to him," Grossman said with a grin.

"I did."

Grossman's eyes lit up. "What did he say?"

"He wasn't happy. He called me a lazy dog."

Grossman bent over in laughter. "A lazy dog," he repeated.

Einstein laughed with him.

"That's a good name for you. I'll have to remember it," kidded Grossman.

"He's lucky I didn't call him a few things," said Einstein, still chuckling.

Einstein and Mileva had stayed in Zurich over the spring break and worked on their theses. They were now complete.

"Three months work," said Einstein holding up his thesis. "I'm handing it in today."

Mileva had handed hers in the day before. Both theses were on the topic of heat conduction. Einstein thought his thesis was adequate, but he knew it contained nothing of importance; with only three months to complete it he was not expected to produce anything of significance. He took the thesis up to Weber after class and handed it to him. Weber opened it and began paging through it. "It's not written on regulation paper," he said.

"Is that important?" Einstein asked. He knew Mileva had written hers on regulation paper but he was sure it didn't matter.

"It's very important," said Weber in a military-like

voice.

"Final exams are only days away. I won't have time to rewrite it."

"You'll have to make time."

Einstein began to panic. "It will take up valuable study time."

"I don't care," said Weber. "I want it written on regulation paper. Maybe you'll learn next time."

Einstein took the thesis back from him. He was late for his next class. Minkowski had announced that he was giving a lecture on capillary and Einstein didn't want to miss it. He sat down just as Minkowski began, not sure what to expect, but pleasantly surprised as the lecture proceeded. He walked out of the room with Louis Kollros. "Well finally," he said, shaking his head. "That's the first lecture on mathematical physics I've had at the Poly."

"You won't be getting any more." said Kollros with a laugh. "The semester is over. All we've got to worry about now is exams."

Einstein was thankful for Grossman's notes, for as usual he had skipped many classes, particularly in mathematics. He was worried as he went into the finals, but he had been forced to study hard because he had got far behind. It was something, he decided, that he never wanted to go through again. Cramming made him lose his enthusiasm for physics.

"I'm worried about math," said Mileva.

"Fiedler?" asked Einstein with a grin.

Mileva nodded.

"Well, just think. This is the last you'll ever have to do with him ... oh happy days." Einstein grabbed her on the shoulder and gave her a little tug.

"It's not just him. I'm worried about Minkowski too. I don't feel I understand his theory of functions well."

"You'll do all right," said Einstein encouragingly. He was worried enough about how he would do to seriously worry about Mileva. Still, he didn't want her to fail.

On the third day of exams Einstein met with Grossman after one of the exams on mathematics. "How do you think you're doing?" asked Grossman.

"They're a lot harder than the intermediates. I'm fairly confident that I'm doing okay in physics, but I'm not so sure about math. Anyway, we'll find out in a few days.

The next few days dragged but finally exams were over and Einstein and Mileva waited nervously for the results. They were to be posted the following day. Einstein was apprehensive as he approached the bulletin board. Mileva was at his side. There was no sign of any of the other students in his class. He wondered where Grossman was. He had likely already seen his grades and left. They stood silently looking at the board. Einstein read the grades, looking only at his and Mileva's.

Experimental physics:	Einstein	10
	Maric	10
Theoretical physics:	Einstein	10
	Maric	9
Astronomy	Einstein	5
	Maric	4
Theory of Functions	Einstein	11
	Maric	5
Thesis	Einstein	18
	Maric	16

Einstein then looked at the totals for all the students in the program:

Kollros 60.0
Grossman 57.5
Ehrat 56.5
Einstein 54
Maric 44

Einstein heard a gasp as Mileva read her total grade. Below were posted the total average grade out of 6. Einstein saw his was 4.91, considerably lower than the 5.7 he had got on the intermediates. Furthermore, he was no longer at the top. The three mathematician's grades ranged from 5.14 to 5.45. Einstein was fourth, with Mileva at the bottom with 4.0. She failed; all the others passed.

Mileva began to sob when she saw that she had failed. Einstein quickly put his arm around her. "It's all right. You can take them again. You'll pass next time."

Mileva continued sobbing on his shoulder. Finally she lifted her head and dried her tears. "This is terrible. It's suppose to be one of the happiest days of my life. I'm engaged and I thought I would be graduating."

"Don't worry," said Einstein, pulling her closer. "Everything will turn out all right."

Mileva finally agreed to retake the exams the following year. Einstein saw her off on the train on the following day.

Pauline and Maja were vacationing at Melchtal, a small resort town south of Lake Lucerne, and Einstein joined them there a few days later.

Maja pulled him to the side as they stepped down from

the carriage. "Walk with me," she said with urgency in her voice.

Einstein lingered back with her as Pauline talked with the driver who was carrying Einstein's suitcases.

"Are you still engaged to Dolly?" Maja asked.

"Yes."

Maja looked at him intensely. "Are you going to marry her?"

"Yes," Einstein repeated nonchalantly.

Maja took a deep breath. "Don't say anything to Mamma. She's strongly against it. Please spare her feelings."

"She has to be told sooner or later," Einstein said stubbornly.

"Maybe later. But please promise me you won't say anything now."

Einstein was silent for several seconds. "I'm not sure I can promise that."

Later in his mother's room he sat down on a chair as she straightened some dressed that were lying across the dresser. "So you passed," she said.

"I didn't do nearly as well as I did on the intermediate exams, but I passed. I ranked fourth ... 4.91 out of 6." He paused. "Now I've got to look for a job."

"What about Dolly?"

He hesitated. "She failed."

Pauline pretended to be surprised. "So what will become of your Dolly now?"

Einstein looked down at the floor. He could feel the tension in the room. In a low voice he said, "My wife."

"Wife!" Pauline said in a loud voice. "Albert, you can't."

"Why not?"

"She's too old for you. She will ruin you," Pauline sobbed. Suddenly she threw herself on the bed and buried her head

in the pillow. Finally she looked up at him. "You are ruining your future and destroying your opportunities. No decent family will have her." She buried herself in the pillow and began sobbing again.

Einstein sat rigid in the chair, uncertain of what to say.

Finally Pauline looked up again. "If she gets pregnant, you'll really be in a mess."

"That's a terrible thing to say. We are not living together." He got up to leave the room when a knock came on the door. Einstein looked over at his mother who was quickly drying her tears on the pillow. She walked to the door and opened it.

"Frau Bar," said Pauline. "Please come in."

"I hope I'm not interrupting anything?" Bar asked innocently. She was a small, vivacious woman, one of the few of Pauline's friends that Einstein liked.

"No," said Pauline. Einstein smiled to himself.

"I'm so glad the weather has cooperated," said Bar. She turned to Albert. "It has been beautiful."

She continued talking to Pauline about the guests in the hotel and how rude some of their children were. "Their parents should keep them under control," she said.

Pauline agreed. "It's almost time for dinner. Would you like to dine with us?" she asked.

Einstein, Maja, Pauline, and Frau Bar ate in the dining room, and after the meal Einstein played the violin. Later, as Einstein was preparing for bed he heard a light knock on his door. He stiffened. He did not want to argue any more with his mother. Hoping it was Maja, he slowly opened it.

"I wanted to say goodnight," said Pauline. She stepped inside.

Einstein felt uncomfortable as she walked in. He said nothing as she sat down in the chair. He sat on the bed.

"Albert, you can't marry her," Pauline began. "It would be a big mistake."

"I'm going to marry her Mamma," Einstein said stubbornly.

Pauline leaned forward. "She's a book, like you. You need a wife. And she's so much older than you. By the time you're thirty, she'll be an old witch."

"I love her."

Pauline closed her eyes and shook her head. "Please, for my sake."

Einstein sat staring at her, saying nothing.

"Have you been intimate with her?" Pauline asked, avoiding his eyes.

Einstein sat up. "No," he said vehemently. "I am shocked you would ask such a thing." He knew that most of what he was doing and saying was for show, but he knew he had to do it.

Finally she got up and kissed him on the cheek. Without another word she left.

Chapter 8

1900 - 1901

Einstein looked out the window at the wet trees and fog. It had been raining for several days and today looked like another day of miserable weather. As he watched the rain hitting the window he wondered if it would ever stop; he had been indoors for several days and ached to get out. He had seen the mountains around Melchtel on the day he arrived; they were lush and green and in the distance were glacier-covered peaks. But since then he had seen nothing but rain and fog. He was still troubled with his mother's reaction to his marriage proposal. Why was she so against it? Why didn't she like Mileva? He wished now that he had kept his mouth shut. But strangely, since her initial outburst she had said little. Maybe she was coming around – resigning herself to the inevitable. She better be, he said to himself, as he was as determined as ever to marry Mileva. She was stubborn, but not as stubborn as he was. He was proud that he hadn't given an inch. It would have been easy to tell her that he had decided that she was right, and Mileva wasn't right for him, but he wouldn't do it.

He had been doing everything he could over the last few days to placate her. He had played the violin for her friends several times, and her face had glowed as the small audience clapped and asked for more. She reveled in his popularity, as it reflected on her and made her feel good. Einstein was happy to do it because it kept her mind off the "Dollie affair" as she called it. Taking a last look as the overcast skies he got up and walked down to the dining room. It was small, but elegantly furnished with small tables and high-backed chairs. A large, draped window extended along one wall. Spotting Maja sitting at one of the tables he sat down beside her. She smiled as she raised a piece of toast to her mouth. She was wearing a dark turtleneck sweater. "Where's Mamma?" he asked.

"She's coming. She's still in her room."

"Has she been saying anything about –" He cleared his throat. "You know who?"

Maja wiped some crumbs off her face as she shook her head. "She hasn't said anything, but I know she hasn't forgotten. I feel the tension whenever I'm around her." She took a deep breath. "I wish this weather would break so we could get out of here."

Einstein smiled as he put some sausages on his plate. "You're not the only one."

Maja leaned back in her chair as she pushed her dark hair back from her forehead. "I'm bored. I wish I were back in Aarau. All Mamma does is sit around and gossip with the old ladies and all you do is study."

"Aren't there any young men around the place?" Einstein said with a chuckle as he cut up his sausage.

Maja frowned. "Have you seen any? I haven't."

"Don't worry. The weather will break."

Maja continued eating her toast in silence. After a few moments she asked, "Have you written to Papa?"

Einstein nodded.

"Did you tell him about your engagement to Dollie? Not that you need to." She rolled her eyes. "I'm sure Mamma has already written to him."

"Yes, I told him, and I know what he'll write back." He knew his father would go along with his mother's objections, but he also knew that his father's feelings would not be as strong. He was sure that if he worked on him long enough, he could bring him around. His mother, however, was another question.

When Einstein finished his breakfast he went back up to his room. It was small and sparsely furnished, but it had a small desk in it. He had moved it near the window where the light was better. He was still studying Kirchhoff, and hoped to finish the book before he left Melchtel. As much as he wanted a break, and wanted to get away, he still loved reading his books. They were an escape. He marveled at Kirchhoff's ingenuity, and continued to be delighted in the knowledge he was getting from the book.

A knock on his door distracted him. He got up and opened it.

"Albert, here's a letter for you," said his mother as she handed it to him. "It's from your father."

He looked down at it, recognizing the writing. His mother stood staring at him, saying nothing. He waited for her to leave but it was obvious she had something to say.

"I wrote your father about the Dollie affair," she said.

"So his letter will no doubt be full of advice," said Einstein, nodding toward it.

"Maybe he can knock some sense into your head."

"Mamma, I don't want to talk about it," Einstein protested.

"I know you don't. But someone has to set you straight. She'll never make a good wife for you. She's unhealthy. She's ... oh what's the use," she cried out in frustration. "I hope you listen to what your father has to say."

"I'd like to read it." He hoped she would go, but he couldn't make it too obvious. Finally she turned and began to leave.

"And pay attention to what he says."

"I will."

As his mother closed the door he sat down on the bed and began to read. As expected the letter was a sermon.

Hermann pointed out that he was in no position to get married, and should wait until he had a good job and had some money saved. Einstein shook his head as he put the letter down.

When Einstein woke the next morning, sunlight was streaming in through his window. He jumped out of bed and ran to it, looking up toward the sky. It was blue with a few scattered clouds on the horizon. His heart jumped. "Finally," he said to himself as he dressed. Moments later he was knocking on Maja's door; she slept in a separate bed in her mother's room, and he knew she wasn't up, but he was anxious to give her the news.

Maja was rubbing her eyes as she opened the door. "Albert," she said, surprised to see him out of bed so early.

"Hurry, get dressed. It's beautiful out. We're going for a hike."

Within minutes Maja was dressed and they were hurrying down the stairs. "We'll take something to eat with us. I don't want to stop for breakfast."

Maja agreed, and they were soon struggling up a trail above Melchtel. Einstein had not hiked for several months and was not in good shape. His legs ached and his lungs felt like they were on fire, but it felt good to be out in the fresh air, away from the chatter and noise of the hotel, so he didn't complain. They finally rested near a large rock and looked down on the small village below them. Maja was breathing hard too, but she smiled as she leaned back. Around them were waves of yellow and white edelweiss, scattered across the mountainsides like stars in the night sky, and in the distance were spectacular mountains with the sun glistening from the snow near their peaks.

"Beautiful," said Einstein as he took a deep breath. The air was fresh and warm, and the only sound was the sound of birds. There were few trees around them, so they could

see for miles. Off to one side was a small brook, its topaz waters trickling over small stones as it cascaded down the mountain to the valley below. The mountains formed a crescent around the valley, rippling with various shades of purple, green and mauve. And above the mountains was a sky as blue as the deepest lake, rimmed to the west with billowy white clouds. A bird flew overhead as the sat marveling at the scenery.

"Did you have another run-in with Mamma?" Maja asked.

Einstein frowned; with all the beautiful scenery around them he didn't want to spoil it with talk about disagreeable things. He looked down at the grass near his feet, then kicked lightly at the ground. "Yes," he said.

"I thought so. She really went at me last night. I don't know why she has to involve me."

"I don't know why she is so stubborn. It's not going to do her any good. We're eventually going to get married anyway." He leaned down and picked up a small stone, rubbed it with his hand, then threw it.

"You got a letter from Papa?"

"Yes." He nodded.

"What did he say?"

"As you might expect, he told me a wife was a luxury that only men with steady jobs and bank accounts could afford."

Maja laughed.

"It puts a wife and a prostitute on the same level," said Einstein,

Maja covered her mouth as her eyes widened. "Albert!

"Well, it's true ... according to them."

They continued up the trail, stopping later to eat the lunch they had brought with them. It was late afternoon when they got back to Melchtel.

That evening Einstein sat down at his desk and wrote

Mileva. He began the letter as he frequently did with a short poem, then went on to tell her about the scene with his mother. He felt guilty about going into so much detail, but she had told him to be truthful with her, and not to hold anything back. A feeling of melancholy and dissatisfaction came over him as he came to the end of the letter. "The people here and their way of life are hopelessly empty," he wrote. "Each meal lasts one hour or more -- you can imagine what hell that is for me."

As Einstein put the letter in an envelope and sealed it he began to wonder about his future. So far he hadn't worried about getting a job, but he would have to start thinking about it soon. He was sure he would get on as an assistant to one of the professors at the Polytechnic. The physics professors needed help in the labs, and the mathematics professors needed graders and help in the classroom. Einstein went through each of the professors in his mind. He dismissed Pernet immediately; he wouldn't want to work for him, and there was little chance Pernet would hire him. Weber was a much better prospect; he hadn't got along well with him, and he was sure Weber didn't have a good opinion of him, but that wasn't important. Aside from Mileva he was the only physics major, and Weber needed help in his lab. He would hire him.

On August 18 Einstein returned to Milan with his mother and Maja. For the last few days at Melchtel his mother had said little about the "Dollie affair," but Einstein was sure he knew why: she was waiting for re-enforcement from his father. Now that they were home he expected them to roll out the big guns and begin their barrage. To his surprise, however, no one said anything during the first two days. He sighed as he thought about it, hoping again that they had resigned themselves.

But the calm didn't last long.

"Albert, come in here. We want to talk to you," his father said after the evening meal on his third day home. They were in the living room. He walked in apprehensively, knowing what was coming. He sat down on a chair opposite them.

His mother began. "We've have been talking about your affair with 'Dollie' and we both feel it would be a serious mistake if you married her."

His father nodded.

"You have a good future ahead of you," continued his mother, "and you need a good wife – a wife that will be an asset to you. As you said yourself, Dolly has several health problems. Why do you want to take on these problems? She is not the right woman for you."

"It would be a mistake son," said his father. "You should seriously reconsider."

"I've already thought seriously about it, and I've made up my mind," said Einstein. He paused. "I love her and I'm going to marry her."

His mother grimaced. "Why are you so stubborn? There are many other girls you could marry."

"She's the one I want to marry. What do you have against her? You've never met her. If you met her, and got to know her, I'm sure you would like her."

Pauline closed her eyes. She knew she would never like her; she didn't even want to meet her. Why was he so obstinate? Maja wasn't like that; she could always bring Maja around to her way of thinking, but Albert – he was impossible! She continued arguing with him for several minutes, then deciding it was hopeless she changed the subject, "You remembered the business trip you're going on with your father?"

"I have to visit the power plants we installed at Cannetto and Isola della Scala," said Hermann. "And I would like for you to come along. You should learn something about the administration of my business, so you could take my place in an emergency. My health is not very good,"

"And we can also visit Venice? Einstein asked.

Hermann nodded.

Einstein enjoyed the trip, but he had his doubts about taking over the business, even in an emergency. Things were not going well for his father, and he was far in debt. Einstein had little interest in going into the business, but he worried about it.

The moment he returned he checked the mail, hoping for a letter from the Polytechnic. He was still optimistic about getting a job, but when he saw there were no letters he began to worry.

Michele Besso was now working for the Society for the Development of the Electrical Industry in Milan. He had married Anna Winteler earlier, and they had a child. Einstein visited him the day after he got back from the trip.

Besso greeted him at the door. "Come in," he said, smiling broadly. Anna was busy in the kitchen preparing the meal. Their house was small, and to Einstein's tastes, strangely decorated; he knew, however, that Anna had done most of the decorating and he never criticized it. Einstein liked Besso but he shook his head every time he thought about him. He was such a wimp – so disorganized, so confused at times, but he had a surprisingly keen mind, and frequently made useful and helpful suggestions. Because of this, Einstein loved talking shop with him.

Anna understood little of what they talked about through the meal, but it didn't bother her. She was used to it; she liked Einstein for his sense of humor, his hearty laughter and genial outgoing personality, and besides ... every time he came over they talked shop. There was little she could do about it, and she accepted it.

After dinner Einstein sat on the couch with Michele. "I'm still bothered by the ether," said Einstein as he rubbed his chin. "I'm convinced it can be done away with."

"You don't think it is needed?"

"It's artificial, and there's absolutely no experimental evidence for it."

"How would electromagnetic waves be propagated without it?"

Einstein nodded his head slowly. "I'm not sure yet, but I have some ideas."

They continued talking about the ether, absolute motion, and the nature of molecular forces for several hours. Einstein was pleased when he left. Besso had given him some good ideas.

The following day Einstein got a letter from the Polytechnic. He looked at it carefully before opening it, knowing that it was in relation to the assistantship he had applied for under Weber. Surely Weber would give it to him, he said to himself as he ripped it open. He read the letter slowly, then closed his eyes. He had not been selected. "Damn," he muttered. He was going back to the Polytechnic in less than a month and had planned on doing a thesis on thermoelectricity under Weber. This put a damper on it. The only other chance of getting on at the Polytechnic now was with the mathematician Hurwitz. Hurwitz wasn't as resentful towards him as Weber, but there was a problem. Hurwitz had been in charge of the mathematics seminar, a seminar Einstein was suppose to have attended, but rarely did, and he was sure Hurwitz would remember. Later that morning he wrote to him.

"Things aren't turning out as I thought," Einstein said to himself as he shaved the next morning. He was shaving regularly now, and although it had been a novelty at first, it was now becoming a tedious but necessary task. The best part of it was that he now had a mustache that he took particular pride in. He smiled as he looked in the mirror. Everything wasn't a disaster; there were bright spots in his life. His parents were nicer to him now, and although he hoped it meant they were resigned to his marriage, he had his doubts. The brightest ray in his life was the thought that he would be seeing Mileva in a few weeks. She was bringing her sister Zorka to Zurich to stay with her, but that was no problem. Furthermore, his work was going well. He was now studying Boltzmann and was even more impressed with him than he had been with Kirchhoff. Boltzmann was at the University of Leipzig and had made many important contributions to thermodynamics and statistics.

The next few weeks dragged but finally Einstein was packing for Zurich. He was excited; it had been over two months since he had seen Mileva and his arms ached for her. He still wasn't sure what he was going to work on for a thesis. Mileva was planning on working under Weber, but Einstein was sure that if he tried to work under him there would be problems. Indeed, after Weber had turned him down, he didn't even want to talk to him. He was still waiting for a reply from Hurwitz, but was confident. If he didn't have anything before he left he would go directly to him and talk to him.

Einstein arrived in Zurich on October 7; Mileva would not be coming until the tenth. Over the next few days Einstein settled in as he waited for her. He was at the station a half hour before her train was due to arrive. It was cold and windy, with dark clouds hovering overhead, but they did not dampen his spirits. He scanned the windows as the train pulled into the station, and although people were waving from many of them, he saw no sign of Mileva. As the train stopped he pushed his way through the crowd toward it. Finally he spotted her; she had already spotted him and was waving. Einstein made his way to her.

Mileva rushed into his arms; she still had one hand in Zorka's hand. After a brief hug she turned to her sister.

"This is Zorka," she said, and to Zorka, "this is my friend Albert, but I call him Johnnie." Zorka had dark hair and a dark complexion like Mileva, but she was slightly shorter. She was eighteen, but seemed younger.

"What am I suppose to call him?" asked Zorka.

"You can call him Albert."

She turned to Einstein. "I'm glad to meet you Albert."

"I had quite a time trying to coax her to come," said Mileva shaking her head. "She was worried she would get homesick."

Einstein looked down at her. "Welcome to Zurich. We'll have so much for you to do you'll never even think of getting homesick. Did your sister tell you we're going to save up and get bicycles so we can go bicycling every Sunday."

"Only on Sundays. What will I do the other days?"

Einstein put his arm around her. "We'll find something for you to do."

They were barely seated in the carriage when Mileva asked about his mother. "Did she really get that upset when you told her about us?"

Einstein nodded his head slowly. "It wasn't a pretty sight." He reached over and grabbed her hand and pressed it. "But don't worry. Everything will work out. We're together now."

"Why doesn't she like me?"

"She thinks you're 'bewitching' me."

Mileva sat back in her seat annoyed, but she was determined not to let his mother's feelings depress her. She looked over at Zorka and forced a smile, then turning back to Einstein she said, "Did I tell you that Helene is getting married in November."

Einstein looked surprised, "No ... to Savic?"

"Yes."

"Poor Helene. He's not much of a catch. She'll suffocate in his fat." He laughed.

"She's very happy."

"That's because she doesn't know what she's in for."

Mileva disagreed strongly but said no more. In many ways she envied Helene. Savic was an engineer and had a good job, and Mileva had never seen Helene happier. She wondered if her and Albert would ever be looking forward to a wedding day in the same way. But right now that wasn't important; the only thing of importance was that she was here with him. "I'm sorry Weber turned you down," she said, "Have you heard anything from Hurwitz yet?"

Einstein let out a deep sigh. "No. I think he's down the drain too, but I'll go and talk to him."

Before he talked to him, however, he got a letter of rejection.

Einstein was disappointed in the small amount of money they were able to make from tutoring. He no longer had his allowance from Aunt Julie and Mileva only had a small amount of money from home, so the money was badly needed. He was still hoping to get an assistantship at the Polytechnic, but the prospects now looked dim. He sat at his desk working on a paper on capillarity; he was planning on submitting it to Allenen der Physik. Although it was based on a simple idea, he was convinced it was completely original and of considerable importance. He had now given up the idea of working under Weber on a thesis, but had talked to Dr. Kleiner at the University of Zurich about his work on molecular forces, and Kleiner had accepted him as a doctoral student.

It was a cold day in December. Snow had fallen all day and Einstein was anxious to get into his warm room. He knocked the snow off his feet as he entered his boarding house. Looking down at the small table inside the door he was delighted to see two letters addressed to him. Picking them up he saw that one was from his parents and the other from the editor of Allenen der Physik. He ripped open the letter from Allenen der Physik first. A feeling of joy overcame over him as he read that his paper had been accepted. This was it – the beginning, he thought. His first paper. He would be able to include a reprint when he applied for jobs. He was sure it would make a difference; he was a published scientist now, and they would have to pay attention. Furthermore, he was overflowing with ideas, and papers would soon be coming in a flood. He couldn't wait to tell Mileva.

He looked at the letter from his parents with apprehension. His father's business had been on the brink for months; he prayed it wouldn't be bad news. He rubbed his chin as he read the letter. His mother was pleading with him to come home for Christmas. Christmas wouldn't be the same without him, she said. And he had to remember that a detective was still checking him out for citizenship; it wouldn't look good if he stayed in Zurich with Dollie over the holidays.

He considered his mother's words. Would it look bad if he stayed in Zurich with Dollie over the holidays? He doubted it; he was here with her now, and he was sure that he would not be denied citizenship for having a girlfriend. They weren't living together, so there was nothing immoral. But it was obvious that his mother wanted him home badly. He thought about Maja and his father; they would be delighted if they knew he was coming.

On the following morning Mileva was waiting for him as usual in front of her boarding house. It was only a short distance from the Polytechnic and they walked to it together. "I've got some good news," Einstein said as he approached her.

Her eyes lit up. "You got a job."

"No, but the paper I submitted to Allenen der Physik was accepted. It will be published in March."

"Oh Johnnie, I'm so happy for you." She gave him a hug, then a quick peck on the lips.

They began walking through the snow. Einstein knew he had to tell her about the letter from his parents. He had thought about what they wrote and had decided it was best to go home for Christmas. He was having a hard time with finances; he had barely been able to come up with December's rent. His dream of a beautiful future with Mileva, unlocking the secrets of nature together was not turning out as he had anticipated. Mileva seemed happy, but money was a continual worry. "I got a letter from my mother too," he said.

Mileva looked at him in dread; she still worried continuously about his parents. Why were they trying to destroy her happiness? Johnnie was happy with her, and she loved him with all her heart. What more could they ask? She didn't understand them. "What did they say?" she asked.

"I don't want you to get mad if I tell you," He grimaced as he looked over at her.

She said nothing as she continued walking, but Einstein could see the tension in her face.

"They want me to come home for Christmas."

Mileva's mouth dropped. "Johnnie, you promised you would stay with me."

"I know. I'm sorry. I would like to, but I should go back to Milan."

"I was already planning everything. We were going to be together." Her face was pale with shock. "You won't come back. I know it. They'll poison your mind against me."

"No, they won't. I'll come back, and we'll continue working together.

Einstein spent Christmas in Milan and was back in Zurich in early January. His parents said nothing about Mileva while he was there, but he knew they were still concerned. He continued working on his thesis under Kleiner and began working on a second paper on capillarity that he would submit to Allenen der Physik. He now knew he wasn't going to get a job in Zurich, particularly at the Polytechnic and he began looking elsewhere, but the prospects continued to be dim. In February he got a letter from the emigration office. The detective's report was complete and he was to come in for another interview. For the first time he began to worry that he might not get citizenship. He was, after all, unemployed, and his parents had a meager income and had declared bankruptcy. He was pleased with himself, however, when the interview was over. A few days later on February 21, he was granted citizenship, but it cost him 200 francs – almost all his savings.

"I'm a Swiss citizen," he said with delight to Mileva that evening.

"Congratulations, we'll have to celebrate. We'll eat out at one of the cafes tonight."

"Can we afford it?" Einstein asked skeptically

"It will be my treat," said Mileva.

Mileva was happy for him but worried. Now that he was a Swiss citizen he was eligible for the draft, and could be inducted into the military. If so he might have to leave Zurich.

The draft notice came a few days before his twenty-second birthday in March. Einstein had hated the army when he was in Munich, and had left Germany to avoid it, but he was now so delighted at becoming a Swiss citizen he looked forward to his military obligation. Still, he hoped he would serve in Zurich, so he could continue with his studies.

"I have to take my military medical exam tomorrow," he said to Mileva the next morning when he met her.

She smiled. "I thought you were through with exams."

She meant it as a pun.

"Well at least this is an exam I don't have to worry about passing."

"Are you sure?"

"Aside from my usual grumbling stomach I'm pretty healthy." He chuckled. "I failed at so many other things. I don't want to fail this."

"Johnnie, you're too hard on yourself," she said.

Einstein sat in the huge office with several other young men waiting for his turn to be called. It had a high ceiling with walls constructed of boards that were vertical and varnished brown, and it was sparsely furnished. Several magazines lay on a small table in front of the row of chairs they sat on. On the opposite side of the room was a long counter. Several young men and a woman sat at desks behind it. Einstein looked down the line of faces waiting to be called. Most looked bored, and one was resting his head against the wall with his eyes shut. One by one they were called into a small adjoining office. Finally Einstein's name was called. He walked into the office.

"Disrobe down to your shorts," said the doctor. "Hang your clothes over there." He pointed to several hooks on the wall. The doctor was middle-aged, bald except for a light brown fringe and he wore thick glasses.

"Step on the scales," he said curtly to Einstein as he finished stripping. Einstein stepped on the scales and waited as the doctor adjusted the weights.

"Over here," he said, pointing to a scale for measuring his height. Einstein moved over to it. "Five feet seven and a half," he said as he recorded the number on his form. He then took a tape and measured his chest, "Thirty-four and a half," he said as he recorded it. Over the next few minutes he checked his knees, feet and other parts of his body. "Dismissed," he said as he handed his report to a younger man. "Check in on the way out."

Einstein put his clothes on and walked out to the counter in the outer office. After checking in he stood waiting as more papers were filled out on him. Finally one of the men approached him; he had a round face and a bushy mustache.

When do I report?" asked Einstein.

The man looked at him quizzically. "You don't have to ... you failed the exam."

Einstein's eyes opened wide in astonishment. "I failed ...what was wrong?"

The man pushed the report in front of him. "You have varicose veins and flat feet, and will not be up to the rigors of the marching that is required."

Einstein was disappointed but when he thought about it he realized it was a relief and he smiled. He would be able to continue working on his thesis and research without interruption.

Mileva was waiting for Einstein in front of her boarding house. The snow was deep and it was cold, but she was bundled in a heavy dark coat, and had a wool shawl wrapped over her head. She smiled as Einstein approached but he knew the smile would not last for long. He had already started to hint to her that things weren't working out. Their meager income from tutoring wasn't enough. They barely had enough to pay the rent, and there was little left over for food. This is not what Einstein had envisioned. Furthermore, his prospects for a job had got no better. He was now the only one in his class, aside from Mileva, that didn't have a job, and it wasn't from lack of trying.

Einstein grabbed Mileva's hand. "Good morning."

She greeted him as she fell in beside him.

"I want to talk to you," said Einstein hesitatingly.

She looked at him apprehensively, knowing what he was

going to say.

"We can't continue living like this."

"It's not that bad Johnnie, and we have each other."

"But we can't eat each other, and I haven't had a decent meal in weeks."

Mileva looked down. "I'm sorry. I've got a little money put aside. I'll treat you tonight."

"No, I think I should return to Milan until I can get on my feet."

Mileva shivered. "Johnnie, we'll make it. As soon as you get a job –"

"That's the problem," Einstein interrupted. "I haven't got a job, and neither have you, and I think I would have better contacts if I were in Milan. There's nothing at the Polytechnic for me."

"What about your thesis?"

Einstein frowned. "I would like to finish it before I leave, but I don't think I can. And I can work on it in Milan."

Mileva was devastated but she tried to keep her composure. "I'll miss you, but I don't want to stand in your way of getting a job."

Einstein squeezed her hand. "I'm sorry, but I think it is for the best."

Einstein arrived back in Milan on March 23. He now had reprints of his article on capillarity and hoped they would be helpful in getting a job. His first letter went to Friedrich Ostwald, a physical chemist at the University of Leipzig. Ostwald's work on capillarity (along with Milkowski's lecture) had inspired Einstein's work. He enclosed a reprint of his article and asked if any positions were available. Subsequent letters went to Göttingen, Stuttgart, Vienna, Bologna and Pisa.

Several weeks later Einstein sat on the sofa in the living room of his parent's house. His father sat across from him on a large soft chair. "Have you heard anything from Dr. Ostwald yet," Hermann asked.

Einstein shook his head. "I even sent him a follow-up letter, saying that I may have forgotten to include my address. But it doesn't appear as if it has done any good."

"What about that other professor?" He hesitated. "Professor Rieche, was it ... in Göttingen?"

"He advertised two assistant posts, but I haven't heard from him either, and I've given up." He took a deep breath and let it out slowly. "I don't understand why they don't even bother to reply. You would think they would have the courtesy of at least telling you that they don't want you."

"Maybe they're too busy."

"I think my problem is Weber." Einstein pursed his lips as he slowly moved his head from side to side. "I've been using him as a reference, and I don't think he's been of any help. I'd like to see what his letters of reference look like."

"Maybe you should write him and encourage him to write better letters of reference."

"I did. I told him he can't continue getting away with it." He paused as he ran his hands through his thick hair. "Maybe I was too harsh. I should have just kept my mouth shut. Anyway I think I'll quit using him as a reference." Einstein looked across at his father. He felt sorry for him; Uncle Rudolf was still pestering him for money, and it was hard on his health.

Hermann stared back at his son. He was troubled that he could do nothing for him. His business was in shatters, and it was only a matter of time. The only thing he could do was write a letter in hopes that it might help. After Einstein had retired that evening Hermann sat down at his small desk and began writing.

Dear Herr Professor Ostwald:

My son, lacking a post at present, feels deeply

unhappy and each day the thought gains strength in him that his career has been derailed and he cannot find a connection any longer. He is moreover depressed at the thought that hc is a burden to us, who are not well off.

As my son probably reveres and esteems you, dear Herr Professor, more than any other physical scientist working today, I am taking the liberty of turning to you with the respectful request you will read the treatise published by him in Allenen der Physik and perhaps send him a few lines of encouragement, so he might regain his vitality and working vigor. If, moreover, you were to find it possible to find for him an assistant's post, either for now or for next Autumn, my gratitude to you would be boundless."

The following day Einstein got a rejection letter from Rieche. As he put the letter down he saw that there was also a letter from Grossman. He was pleased; he hadn't heard from him in a while and was always happy to get a letter from him. He felt guilty, however, as he had been pouring out his troubles to him, and he was starting to become embarrassed about it. He opened Grossman's letter and began reading. A warm feeling came over him as the words were absorbed. Grossman had talked to his father about his friend's difficulties, and his father in turn had talked to Friedrich Haller, the director of the patent office in Bern. A position would be coming up in the patent office, and if he applied for it there was a good chance he would get it. Einstein was overjoyed. He sat down and wrote Grossman immediately. "I am truly touched by your loyalty and humanity which did not let you forget your old luckless friend ... I need hardly tell you that I would be happy to be granted such a fine field of activity, and I would do everything not to disgrace your recommendation."

Einstein was walking on air for the rest of the day. There would be a waiting period of a few months, but at least he had a prospect. It was like a ray of light through a heavily overcast sky and he was delighted. The following day he received another letter; it was postmarked Winterthur. He tried to think who he knew in Winterthur as he opened the letter. It was from Jacob Rebstein at the Technical high School. Einstein's heart pounded as he read the letter. Rebstein had a military obligation from mid-May until July. Would Einstein like to take over his teaching position during this time? Einstein brought up his arms in delight. His luck was changing. He would write Mileva immediately. They had to celebrate. "Come to Lake Como," he wrote.

A few days later he got a reply. "I received a letter from home today and it has made me lose all desire, not only for having fun, but for life itself. Don't let it bother you though, go ahead and take the trip..."

Einstein could not believe what he was reading. He was so happy; how could she be so gloomy. He wrote her again. "I just won't let up. You absolutely must come to see me in Como, you sweet little witch. It would cost very little of your time and will he a heavenly joy to me ..."

Mileva relented. She arrived at Como before he did and was waiting for him. "My little kitten," he said as she ran into his arms. "I'm a new man. No more sour puss around here." He took a deep breath of the air, then looking around at the people and buildings he smiled. "What do you want to do first?"

"I just want to be with you," said Mileva, looking at him admiringly.

They wandered down to the main promenade and look over the lake. The view was breathtaking, with glacier-covered mountains in the distance, and steep cliffs surrounding the lake in many places. And the sights in the city itself were almost as entrancing: Piazza Duomo with Torre del Comuno and il Broletto with its white marble-striped parapet and the large Gothic-Renaissance Cathedral with the smaller temple containing the pagan statues of Pliny the Elder and his nephew, and in the distance the Temple of Sant Abbondio. Finally they stopped at a cafe. Einstein ordered coffee and a pastry; Mileva ordered only coffee. They looked out over the lake as they sipped their coffee. "Let's take the boat," said Einstein.

Mileva looked at him in surprise. "Can we afford it?"

"We're rich. Have you forgot. I've got a job coming up."

They boarded the boat to Cadenabbia just after lunch. As they stood on the deck they looked down at the blue clear water of Lake Como. Mileva hugged Einstein; she was happy. The boat stopped at Villa Carlotta. "Let's tour the gardens," said Mileva. Einstein agreed. They started with the villa's collection of marble statues by Canova, then strolled into the lush terraced gardens where they saw azaleas, camellias and rhododendrons and other types of flowers.

"All this beauty and we can't pick even a single flower," said Mileva with a laugh.

When they finished touring the gardens they returned to the boat. It was back in Como by five o'clock. They had reserved two rooms earlier and went to them briefly before eating dinner.

After delightful evening visiting cafes along the waterfront they returned to their rooms. Mileva had barely got into bed when she heard a light tap on her door. She got up and went over to it. "Johnnie?"

"Yes, let me in."

Mileva opened the door.

"I just came to say goodnight." He stepped inside. "Were you in bed?"

She nodded.

They walked over and sat on the edge of the bed. He leaned over and kissed her as he caressed her hair.

"I don't think we should Johnnie. What would your

mother say if she found out?"

"How could she ever find out?"

"I could get pregnant."

"I don't think there's much chance of that." He leaned back in the bed. She curled up beside him.

The following day Einstein grinned happily as the carriage carried them higher and higher into the mountains. He looked out the window at the spectacular scenery around them: steep ravines, cascading waterfalls interspersed with lush fields of green. They were heading for Splugen Pass. Mileva cuddled as close to him as she could get. As the carriage wound up bend after bend the air got colder, but they were huddled so close together they hardly noticed.

The sky was a dull gray, and soon the first flakes of snow began to snow. Mileva smiled as Einstein pointed them out to her. "Maybe we'll get stranded and have to stay here," she said jokingly. But soon it was snowing heavily, and the carriage pulled into a lodge.

"This is the end of the line," the driver shouted down to them.

They got out and looked around. There was considerable snow on the road ahead. "If you want to go any further you'll have to hire a sleigh," the driver said to them.

Einstein looked over at Mileva with a grin on his face. "That's a good idea." He spotted some sleighs parked near the lodge and headed for them. Twenty minutes later they were again huddled together on a small sleigh. The driver stood on a small plank on the back, holding the reins of the horse that pulled them. It was snowing even heavier now. They pulled themselves under the heavy blanket that was draped over them as the driver chatted away amiably. The snow got heavier until they could see nothing but a field of white. "I'm glad the driver knows where he is going," said Einstein. "I'm lost."

"What if he's lost too?" said Mileva, laughing.

"Then I guess we'll just have to build ourselves a snowcave." He winked at her.

The driver finally halted the sleigh. "This is as far as I can go," he said.

Einstein and Mileva got out from under the blankets and began walking. It was still snowing heavily and was now getting dark. The laughed as they threw snowballs at one another. Suddenly the snow stopped and they could see tiny houses in the valley below them. They stood looking at them for several minutes, then returned to the sleigh. The following day Einstein left for Winterthur.

Einstein was happy with his job at Winterthur. Although the teaching schedule was rather heavy, it left him time to study. He met an old friend, Hans Wohlwend who told him there was a vacant room at the boardinghouse where he stayed. Einstein took the room and later joined the local orchestra with Wohlwend. Each Sunday he traveled to Zurich to be with Mileva.

He had only been working a few weeks when he got a letter from his father asking if he could help pay for Maja's education in Aarau. The business was on the verge of bankruptcy. Einstein sighed. He had been expecting it, and knew it would be difficult to save 30 francs a month for Maja, but he would do it.

He met Mileva the following Sunday. She looked forlorn as she met the train. "What's wrong?" he asked. "No smile for Johnnie?"

"I'm sorry." She gave him a forced smile.

Einstein put his arm around her and they began walking. Mileva kept her head down.

"Something's wrong, isn't it?" Einstein asked.

Mileva pulled closer and looked up at him. "You won't be mad if I tell you?"

Einstein looked at her in surprise. "Me, I never get mad."

"I'm pregnant."

To be continued in **Young Einstein II**

Barry Parker is a professor emeritus at Idaho State University where he taught physics and astronomy for 30 years. He is the author of 25 books and has written for the Smithsonian, Encyclopedia Britannica, Time-Life Books, the Washington Post and numerous magazines including *Astronomy, Sky and Telescope, flyfisherman* and *Fishing World.* He taught a writing class at ISU for several years, and has recently self-published several books. For more information see his webpage at BarryParkerbooks.com. His email is parkbarr@isu.edu

Additional copies of "Young Einstein" can be obtained from Amazon.com or Barnesand Noble.com, or directly from the author at BarryParkerbooks.com

LaVergne, TN USA
17 January 2010
170265LV00001B/1/P